PEPTIDE AND PROTEIN REVIEWS

PEPTIDE AND PROTEIN REVIEWS
Volume 4

Executive Editor
MILTON T. W. HEARN

ST. VINCENT'S SCHOOL OF MEDICAL RESEARCH
VICTORIA PARADE
MELBOURNE, VICTORIA
AUSTRALIA

MARCEL DEKKER, INC.   New York and Basel

MARCEL DEKKER, INC.
270 Madison Avenue, New York, New York 10016

ISBN 0-8247-7292-X

Current printing (last digit):
10   9   8   7   6   5   4   3   2   1

PRINTED IN THE UNITED STATES OF AMERICA

PREFACE

Peptide and Protein Reviews is a comprehensive series of volumes
devoted to the techniques, concepts, and practice of peptide and
protein chemistry in biological research. The significant expan-
sion of biomedical research, and life science in general, over the
past decade has led to the progressive subdivision of peptide and
protein chemistry into highly specialized areas. The volumes in
the Peptide and Protein Reviews series - of which this is the
fourth - serve as a focus among basic disciplines involved with
methodological advances and the ensuing biological areas of investigation that benefit from these developments. Previous volumes in this series have dealt with such topics as new methods
for the isolation, characterization, and synthesis of peptides;
biosynthetic pathways involved with the expression and modulation
of peptide hormones; solution conformation of enzymes, and other
homologous proteins; and organization of membrane receptors and
polypeptide microsequencing.

Over the past decade the hierarchy in structure and organization of proteins as revealed in particular by X-ray crystallographic techniques has become increasingly evident as the role of folding units, domain structure, and protomer-oligomer assembly has been
progressively identified. This volume deals with the structure-
function relationships of a variety of biologically important enzymes as revealed by X-ray and neutron diffraction studies and
associated protein crystallographic investigations. Important
recent discoveries in this field are reviewed by leading author-
ities.

This volume should thus be of utmost interest to all scientists working in the area of protein structure and organization,
protein crystallography and enzymology as well as in general to
medical and life scientists. We trust that you will make this and
other volumes in the Peptide and Protein Reviews series an integral
part of your regular professional reading, in your undergraduate
and graduate course preparation and, most importantly, as a useful
adjunct to the exchange of ideas with your scientific colleagues.

                                        Milton T.W. Hearn,
                                         Executive Editor

CONTENTS

CONTRIBUTORS

ARTHUR ARNONE   *Department of Biochemistry, University of Iowa, Iowa City, Iowa*

LEONARD J. BANASZAK   *Department of Biological Chemistry, Washington University School of Medicine, St. Louis, Missouri*

JENS J. BIRKTOFT   *Department of Biological Chemistry, Washington University School of Medicine, St. Louis, Missouri*

PATRICK D. BRILEY*   *Department of Biochemistry, University of Iowa, Iowa City, Iowa*

PETER M. COLMAN   *CSIRO Division of Protein Chemistry, Parkville, Victoria, Australia*

BARRY C. FINZEL†   *Department of Chemistry, University of California, San Diego, La Jolla, California*

C. CRAIG HYDE   *Department of Biochemistry, University of Iowa, Iowa City, Iowa*

RUDOLF LADENSTEIN   *Abteilung Strukturforschung II, Max-Planck-Institut fuer Biochemie, Martinsreid, West Germany*

CAROL M. METZLER   *Department of Biochemistry and Biophysics, Iowa State University, Ames, Iowa*

DAVID E. METZLER   *Department of Biochemistry and Biophysics, Iowa State University, Ames, Iowa*

THOMAS L. POULOS†   *Department of Chemistry, University of California, San Diego, La Jolla, California*

PAUL H. ROGERS   *Department of Biochemistry, University of Iowa, Iowa City, Iowa*

*Current affiliation: Digital Equipment Corporation, Wheeling, Illinois
†Current affiliation: Protein Engineering Division, Genex Corporation, Gaithersburg, Maryland

PEPTIDE AND PROTEIN REVIEWS

STRUCTURE-FUNCTION RELATIONSHIPS AMONG
NICOTINAMIDE-ADENINE DINUCLEOTIDE DEPENDENT OXIDOREDUCTASES

Jens J. Birktoft and Leonard J. Banaszak
Department of Biological Chemistry
Washington University School of Medicine
St. Louis, Missouri

ABSTRACT

The known crystal structures of the nicotinamide-adenine dinucleotide dependent oxidoreductases have been analyzed and compared. The dinucleotide binding domains in all of these enzymes
contain a common structural motif that consists of a four stranded parallel β-sheet and one α-helix. The same structural motif
is also found in the domains that bind the dinucleotide FAD. In
all instances the topography is the same, +1, +1, -3, with the
helix forming the connection between the first two β-strands and
structurally equivalent residues performing similar roles in
enzyme-dinucleotide interactions. Several invariant glycine
residues permit the close proximity of the two ribose rings to
the β-sheet and of the pyrophosphate to the α-helix. An acidic
group interacts with the adenine ribose 2-hydroxyl groups in NAD
and FAD. A comparison of the dinucleotide binding sites in terms
of the nicotinamide ring is made for this group of enzymes.

Those enzymes that utilize the nicotinamide-adenine di-

nucleotide NAD and NADP, and their reduced counterparts NADH and

NADPH as cofactors, form a large and diverse group of proteins.

More than 250 such enzymes have been identified (1). A variety

1

of names have been used for these enzymes, but there appears to
be little if any systematic pattern in the resulting nomenclature. Thus, names such as dehydrogenase, reductase, reducto-
isomerase, oxidase, transhydrogenase, hydroxylase and monooxygenase are among those that have been employed. Oxidoreductase
is the recommended generic name for enzymes involved in oxidation
reduction reactions in general, but neither this nor any of the
other names which have been used do necessarily imply that NAD or
NADP participates in the redox reaction.

     Despite several attempts to define unifying features among
the NAD/NADP requiring oxidoreductases, few if any seem to exist
(2). The only common denominator appears to be their ability to
catalyze the transfer of a hydride ion to and from the C-4 carbon
atom of the nicotinamide ring. Most of these enzymes display
strict specificity towards either NAD or NADP, whereas a more
limited number of oxidoreductases are able to utilize either coenzyme. In the latter instance, the enzymes involved have different affinities toward the two dinucleotides. The two hydrogen
atoms located at the C-4 carbon in the dihydronicotinamide ring
are not enzymatically equivalent. This aspect of stereospecificity appears to be a unique and characteristic feature of a given
enzyme in that the transfer of a hydride ion always takes place
to the same side of the nicotinamide ring (2,3).

     The term, "hydride transfer", will be used repeatedly in the
text that follows. The reader should remember that this term is
used to describe the transfer of one proton and two electrons between the coenzyme and substrate, and not the true catalytic
mechanism for the reaction. The type of reaction catalyzed by
NAD/NADP dependent enzymes covers a range varying from simple,
direct hydride transfers between the nicotinamide ring and a
substrate to enzymes where the redox reaction is only one step in
a more complex sequence of chemical reactions. To illustrate
these situations with examples for which detailed structural

information exists, malate dehydrogenase and liver alcohol dehydrogenase would exemplify the former type and glyceraldehyde-3-
phosphate dehydrogenase (phosphorylation of substrate) and p-
hydroxybenzoate hydroxylase (incorporation of molecular oxygen)
the latter. As another example of the diversity among the NAD
and NADP dependent oxidoreductases, it should be noted that in
some instances the transfer of a hydride ion occurs directly
between the substrate and the nicotinamide ring, whereas in other
systems another cofactor or prosthetic group acts as an intermediate. In the discussion that follows, we shall call these simple
and complex NAD/NADP requiring oxidoreductases respectively. The
nicotinamide-adenine dinucleotide is fully exchangeable with
solvent in both types of oxidoreductases, whereas in general, the
other cofactor in the complex oxidoreductases is not.

Considering the large number of known oxidoreductases, relatively few have been subjected to study by x-ray crystallographic
methods. In Tables 1 and 2 are listed the enzymes for which
structural analyses have been done at a resolution which permits a
tracing of the polypeptide chain. In some instances accompanying
studies have been carried out on binary and ternary complexes
formed with coenzymes, coenzyme analogs, substrate, substrate
analogs and inhibitors. Note that Tables 1 and 2 also contain the
abbreviations for each protein which will be used in the following
text. Generally, detailed structural information which is now
available on the interactions between these enzymes and their
coenzymes has helped to establish functional and evolutionary
relationships (4).

Structural information concerning the binding modes of
substrates and more critically on the structures of productive
ternary complexes is much more limited. Investigation by x-ray
crystallography of productive enzyme-substrate complexes is normally not possible unless special low temperature methods are
employed (5). Therefore, much of our current ideas concerning the

TABLE 1

High Resolution X-ray Investigations of NAD Dependent
Oxidoreductases

Enzyme	Resolution	References
Malate Dehydrogenase, (sMDH), pig heart cytoplasm		
Apo	4.5 Å	6,7
+ NAD	2.5 Å	8,9
Lactate Dehydrogenase, (LDH), dogfish muscle		
Apo	2.0 Å	10
+ ADP	2.8 Å	11
+ citrate	2.8 Å	12
+ NAD + pyruvate	3.0 Å	13
+ NAD + oxalate	3.0 Å	13
+ NADH + oxamate	3.0 Å	13
LDH, pig heart		
+ S-lac-NAD*	2.7 Å	14
LDH, mouse testes		
Apo	2.9 Å	15
Glyceraldehyde-3-Phosphate Dehydrogenase, (GAPDH), lobster		
Apo	2.9 Å	16
+ NAD	2.9 Å	17,19
+ 8-bromo NAD	2.9 Å	19
+ NAD + citrate	2.9 Å	18
+ trifluoroacetone	2.9 Å	20
GAPDH, B. Stearothermophilus		
Apo	2.5 Å	21
+ NAD	2.7 Å	22
Liver Alcohol Dehydrogenase, (LADH), horse		
Apo	2.4 Å	23
+ ADP-ribose	2.9 Å	24
+ imidazole	2.9 Å	25
+ NADH + DMSO*	2.9 Å	26
+ H$_2$NADH* + DACA*	2.9 Å	27
+ H$_2$NADH + MPD	2.9 Å	27
+ NADH + pyrazole	2.9 Å	28
+ NAD + 4-iodopyrazole	2.9 Å	28

*Non-standard abbreviations are listed in Table 2.

TABLE 2

High Resolution X-ray Investigations of NADP

Dependent Oxidoreductases

Enzyme	Resolution	References
Dihydrofolate Reductase, (DHFR), L. casei		
+ NADPH + methotrexate	1.7 Å	41,42
DHFR, E. coli		
+ methotrexate	1.7 Å	41
DHFR, chicken liver		
+ NADPH	2.9 Å	43
+ NADPH + phenyltriazine	2.9 Å	43
Glutathione Reductase, (GTHR), human erythrocytes		
+ FAD	2.0 Å	44,45
+ FAD + NADPH	3.0 Å	46
+ FAD + NADP + GSSG*	3.0 Å	46
+ FAD + NADP	3.0 Å	46
+ FAD + GSH*	3.0 Å	46
p-Hydroxybenzoate Hydroxylase, (PHBH), Pseudomonas fluorescens		
+ FAD	2.5 Å	48
+ FAD + p-hydroxybenzoate	2.5 Å	48

*Non-standard abbreviations: S-lac-NAD, (3S)-5-(3-carboxy-3-hydroxypropyl)-NAD+; DMSO, dimethyl sulfoxide; H_2NADH, 1,4,5,6 tetrahydronicotinamide adenine dinucleotide; DACA, trans-4-(N,N-dimethylamino)cinnamaldehyde; MPD, 2-methyl-2,4-pentane diol; GSH, reduced glutathione; GSSG, oxidized glutathione.

possible structures of productive complexes is derived from model building studies and from correlations of the crystallographic results with those obtained through other methods such as enzyme kinetics, NMR, isotope effects, etc. In the text which follows, an attempt has been made to summarize the crystallographic studies. In addition, a number of unifying features among these proteins have begun to appear and will be described.

THE STRUCTURE OF NAD DEPENDENT OXIDOREDUCTASES

The majority of the structural work performed on NAD dependent oxidoreductases has focused on just four different enzymes, LDH, sMDH, LADH and GAPDH. These are all simple NAD requiring oxidoreductases in that the hydride ion transfer takes place directly between the nicotinamide ring and the substrate. However GAPDH differs from the other three enzymes in that a covalent enzyme-substrate complex is formed, and it is between the covalently bound substrate and the coenzyme that hydride transfer takes place. Detailed descriptions and comparisons of the structures of these enzymes and their complexes have been presented in several articles and reviews (29-32) and therefore only a brief summary of the structural aspects of the NAD binding domain and the nature of NAD-enzyme interactions will be given here.

The initial comparisons of LDH, sMDH, LADH and GAPDH revealed some striking structure-function relationships. These enzymes are all oligomeric, sMDH and LADH being dimeric, and LDH and GAPDH tetrameric. The subunits of all four dehydrogenases can be divided into two structurally different domains each of which serves a different biochemical function. One domain, the so-called nucleotide binding domain, is involved in anchoring the coenzyme whereas the other provides the amino acid residues involved in substrate binding and catalysis, and is often referred to as the "catalytic" domain. The x-ray studies showed that the NAD binding domain in all four enzymes are highly homologous to each other and that the binding of NAD takes place in a nearly identical fashion. However, in most cases the catalytic domains show little similarity to each other. Only in the case of the LDH-sMDH system was it apparent that the two enzymes were homologous in both the nucleotide binding and catalytic domains.

The central structure of the NAD binding domain in these four enzymes is a six-stranded parallel β-sheet with the connectivity between the β-strands all being right-handed (33) and

in most instances α-helices. The overall arrangement of the
polypeptide chain is virtually identical in the four dehydrogenases, as seen both in the great similarity in the polypeptide
backbone hydrogen bonding pattern (7,34) and in the fact that
this β-sheet in all four enzymes has the same right-handed twist
of approximately 100°. The connecting helices are nearly all
parallel to each other and are approximately antiparallel to the
strands of the β-pleated sheet. The binding site for the NAD is
located at the amino terminal end of these helices and at the
carboxyl terminal end of the β-strands. Stereodiagrams of the α-
carbon atoms of the NAD binding domains of these four enzymes are
shown in Figures 1A-1D. Diagrammatic representations of the dinucleotide binding domains are shown in Figure 2. In Figure 2,
symbols are used to describe the individual elements of secondary
structure. Thus the Greek letter prefixes α and β are used to
indicate alpha helices and beta structures, respectively.   The
second letter describes the relative position in the primary
structure; hence βA is the first strand of extended polypeptide
chain beginning from the N-terminus, βB the second strand, etc.
This nomenclature has been commonly used in the past to describe
the structure of the simple dehydrogenases (34).

    The greatest degree of structural homology among the NAD
binding domains is found in the centrally located β-sheet,
whereas the secondary structural elements located on the periphery of this domain display a much greater variability. For
example the helix, αC of GAPDH (Fig. 1D) differs significantly in
its location relative to the β-sheet when compared with the other
NAD binding domains. The insertion of about 20 additional residues, between αC and βC might well be the reason for this difference. In GAPDH, the additional residues together with βC form
a three-stranded antiparallel β-sheet. This feature can be seen
at the left side of Figure 1D. The connection between βE and βF
at the other end of the β-sheet also displays great variability.
In LDH and sMDH a helix, αF, makes this connection whereas LADH
and GADPH both have rather irregular folded polypeptide chains.

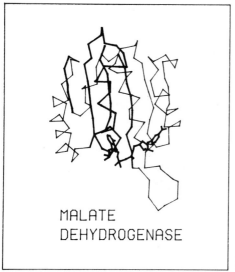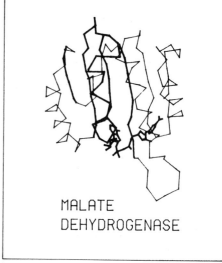

(A)

Fig. 1: α-Carbon models of NAD and NADP dependent oxidoreductases. The enzymes are shown in stereo as α-carbon models
of their polypeptide chains. The bound dinucleotides are indicated
by the heavier lines as is the structural motif containing four
β-strands and one α-helix, that is common to all the dinucleotide
binding proteins. The structures were aligned on a MMS-X graphics
system such that this common structural motif appears in approximately the same relative orientation. The atomic coordinates
used are from the Protein Data Bank at Brookhaven National
Laboratory.

(A) Cytoplasmic malate dehydrogenase (36). Residues 5-149
and NAD are shown.

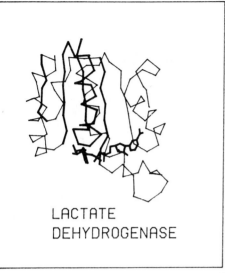 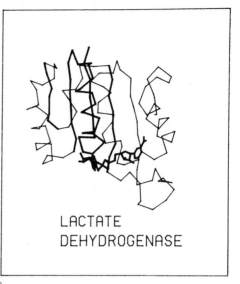

LACTATE
DEHYDROGENASE

LACTATE
DEHYDROGENASE

(B)

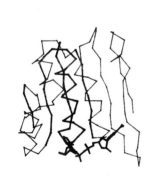 

LIVER ALCOHOL
DEHYDROGENASE

LIVER ALCOHOL
DEHYDROGENASE

(C)

(B) Pig heart lactate dehydrogenase (14). Residues 22–163 and NAD are shown.

(C) Horse liver alcohol dehydrogenase (23). Residues 193–318. The coenzyme shown is H_2NADH and is from (27).

GLYCERALDEHYDE
3-PHOSPHATE
DEHYDROGENASE

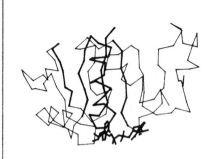

GLYCERALDEHYDE
3-PHOSPHATE
DEHYDROGENASE

(D)

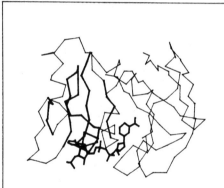

DIHYDROFOLATE
REDUCTASE

DIHYDROFOLATE
REDUCTASE

(E)

Fig. 1 (continued):

(D) Lobster glyceraldehyde-3-phosphate dehydrogenase (17). Residues 1-147 and NAD are shown.

(E) L. casei dehydrofolate reductase (41). Residues 1-162 and NADPH are shown.

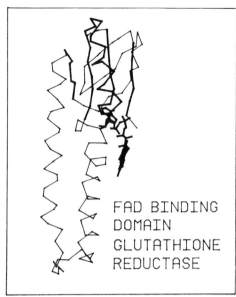
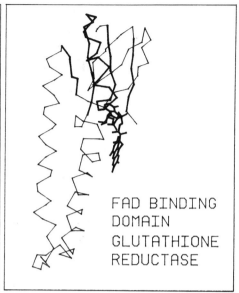

FAD BINDING
DOMAIN
GLUTATHIONE
REDUCTASE

(F)

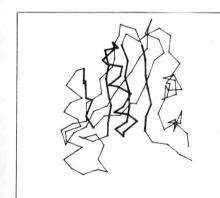

NADP BINDING
DOMAIN
GLUTATHIONE
REDUCTASE

(G)

(F) Human erythrocyte glutathione reductase (44). Residues 19-157 and FAD are shown.

(G) Human erythrocyte glutathione reductase (44). Residues 158-293 are shown.

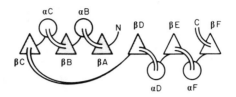

MALATE DEHYDROGENASE

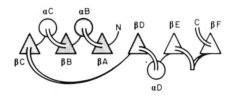

LIVER ALCOHOL DEHYDROGENASE

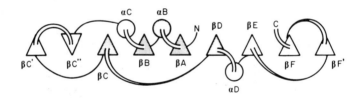

GLYCERALDEHYDE 3 − PHOSPHATE DEHYDROGENASE

Fig. 2: Schematic diagrams of the secondary structure and
folding of the dinucleotide binding domains. Each β-strand is
represented by a triangle, with its apex pointing up or down
depending on whether the C-terminal or N-terminal end is nearer
the viewer respectively.   The circles represent α-helices. Connections occurring at the end of the β-sheets nearest to viewer
are indicated by the thicker lines. The topological diagrams are
related to the stereo diagrams in Figures 1A-G in that the
leftmost β-strands in Figure 1 are also the leftmost β-strands in
Figure 2. The C-terminal end of the parallel β-sheets are at the
lower end of Figures 1A-1G, and nearest to the viewer in Figure

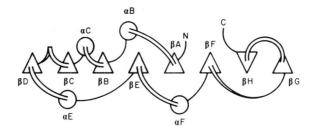

DIHYDROFOLATE REDUCTASE

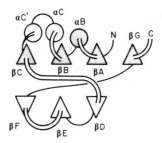

GLUTATHIONE REDUCTASE (FAD DOMAIN)

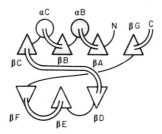

GLUTATHIONE REDUCTASE (NADP DOMAIN)

2. The nomenclature used to describe the secondary structure is explained in the text. The stipling indicates the four β-strand - one α-helix unit that is common to all dinucleotide binding domains. The diagram for LDH is the same as that shown for sMDH.

The segments of polypeptide chain that join the elements of secondary structure vary in both conformation and in length. In LDH and sMDH, an additional 20 residues located between βD and αD form the so-called "NAD-loop" (8,35), located at the bottom of Figures 1A and 1B. In LDH this loop undergoes a conformational change upon binding of NAD and it has been suggested that this movement is important in the binding process (35). It has also been shown that in the most homologous pair of these four enzymes, LDH and sMDH, differences in the length of polypeptide chain occur only in the polypeptide segments connecting elements of secondary structure (36).

Based primarily on their analysis of LDH, Rao and Rossmann (37) proposed that the NAD binding domain could be divided into two similar subdomains. These two super secondary structures or mononucleotide binding units are related by approximately two-fold rotational symmetry. The location of this pseudo two-fold axis would be located between β-strands βA and βD in the top panel of Figure 2.

Despite the extensive structural homology observed in the NAD binding domain, little significant amino acid homology is observed among the four dehydrogenases. The alignment of equivalent amino acid residues in the NAD binding domains worked out by Ohlsson et al. (30) shows that just five out of 94 structurally equivalent amino acids are conserved in LDH, LADH and GAPDH. When this comparison is extended to include the five isozymes of LDH, mMDH and sMDH only four invariant residues remain (38,39). Using the numbering scheme for dogfish LDH (38) these residues are Gly 28, Gly 33, Asp 53 and Gly 99. Three of these residues participate in coenzyme binding, as will be discussed below, whereas Gly 33 appears to be conserved for structural reason. Any larger side chain at this position would perturb the contact made between helix αB and β-strand βA. The contact regions between the helices and the β-sheet show a high degree of con-

servation of amino acid type, and in particular of valines and
isoleucines. Rather than being unique for nucleotide binding
proteins, this feature appears to be characteristic for the
contact regions between α-helices and β-pleated sheets in general
(40). Finally amongst the four simple dehydrogenases, only in the
case of the pair, LDH-sMDH, can any similarities be found in the
so-called catalytic domains (34,36).

THE STRUCTURE OF NADP REQUIRING OXIDOREDUCTASES

Both simple and complex NADP requiring oxidoreductases are
represented among those enzymes for which high resolution structures have been reported (Table 2). Among the simple oxidoreductases the structures of dihydrofolate reductase from three different species have been determined (41-43). The structure of
one other simple NADP requiring dehydrogenase, 6-phosphogluconate
dehydrogenase, has been determined at 6 Å resolution (49), but the
polypeptide chain conformation is yet unknown.

The complex NADP requiring oxidoreductases for which high
resolution structures have been reported include glutathione
reductase (45) and p-hydroxybenzoate hydroxylase (48) and the
structure of a third enzyme, ferredoxin-NADP$^+$ oxidoreductase, is
known at 3.7 Å resolution (50). This group of NADP requiring
oxidoreductases are called "complex" because the oxidation-reduc-
tion reaction which they catalyze involve the movement of electrons from the reduced dinucleotide, NADH or NADPH, through a
second cofactor, usually a tightly bound FAD moiety. The latter,
itself a dinucleotide, is not exchanged through the solvent
during a catalytic cycle and is best described as a prosthetic
group. All three enzymes require NADP as a cofactor. However,
lipoamide dehydrogenase, which has extensive homology with glutathione reductase at the amino acid sequence level, requires NAD
as a cofactor (51). Analysis of these structures indicate that
little conformational similarity exists between the complex and

simple NADP dependent oxidoreductases and consequently they will
be discussed separately.

The Structure of a Simple NADP Requiring Oxidoreductase

The structures of DHFR have been determined for the enzyme
from avian liver (43), Lactobacillus casei (41) and Escherichia
coli (41). In fact, the last two studies have resulted in the
most accurately determined oxidoreductase structures to date.
The conformations of the three forms of DHFR are very similar
with the major differences occurring between the bacterial and
eukaryotic forms of the enzyme. The conformational differences
generally appear in external loop regions of the molecule, regions
which are thought to be most able to accommodate insertional or
deletional differences amongst homologous enzymes (43).

The first question to be asked about the molecular structure
of DHFR is whether or not it contains the nucleotide binding domains reported in the simple NAD requiring oxidoreductases. This
domain would contain the six-stranded parallel β-sheet structure
which has been described in the earlier section. As can be seen
in the schematic diagrams contained in Figure 2, only very crude
structural similarities are to be found between DHFR and the
simple NAD dependent oxidoreductases (52). For example, the
principal conformational similarity found between DHFR and sMDH
is the central core of the dinucleotide binding region which
consists of a multistranded β-sheet structure predominantly of
the parallel type. This β-sheet is flanked on both sides with α-
helices which are roughly antiparallel to the strands in the β-
sheet. The conformation of DHFR from L. casei along with the
bound coenzyme NADPH is shown in the stereoview given in Figure
1E (41,42). To permit easy comparisons, the orientation of DHFR
used in Figure 1E has been adjusted so that the central β-sheet
structure and the adenosine moiety of the bound coenzyme is in
roughly the same orientation as the corresponding features in

sMDH shown in Figure 1A. Note that the entire molecule of DHFR
which contains about 160 residues is depicted. Hence, the
entire DHFR molecule must be compared with only the NAD binding
domain of the simple NAD requiring oxidoreductases, which consist
of roughly one-half of these molecules. Any reference to the
dinucleotide binding domain in DHFR therefore includes the entire
molecule.

A comparison of the topology of the core β-sheet structure
of DHFR, and the simple NAD dependent oxidoreductases indicates
that there are as many differences as there are similarities. As
can be seen in Figures 1 and 2, the connectivity between the
strands in the β-sheet found in DHFR, although all of the right
handed type, are generally different from those in the simple NAD
dependent oxidoreductases. If the labeling of the β-strands is
done following the same convention as employed for sMDH, adjacent
segments of β-structure would appear in the following order as
shown below to the left for DHFR and to the right for sMDH.

$$\underset{\begin{array}{c}\uparrow\\ \text{C-term}\end{array}}{\overset{\text{N-term}}{\beta D - \beta C - \beta B - \beta E - \beta A - \beta F - \beta aH - \beta G}} \qquad \underset{\begin{array}{c}\uparrow\\ \text{to catalytic domain}\end{array}}{\overset{\text{N-term}}{\beta C - \beta B - \beta A - \beta D - \beta E - \beta F}}$$

The symbol βaH is used to define the antiparallel strand of the
sheet structure found in DHFR (41,52). Notice in Figure 2, that
strands βB, βC, βD and βE in DHFR and βA, βB, βC and βD in sMDH
have the same relative connectivity.

As will be discussed in more detail later, this combination
of β-strands may represent the homologous regions among the NAD
and NADP requiring enzymes, and the similarities extend to the
level of coenzyme-enzyme interactions. For example in DHFR, the
adenine ribose moiety of NADPH is close to the carboxyl terminal
end of the β-strand, βB. In sMDH and the other simple NAD dependent oxidoreductases it is the carboxy-terminus of β-strand βA
which is near the adenine ribose. The conclusion reached by

Matthews and his coworkers from the comparison studies of DHFR
and LDH is that the folding of these dinucleotide binding proteins is only superficially similar (52).   Only when a subset of
the β-sheet structure is considered can some similarities between
these enzymes be recognized.

The Structure of Complex NADP Requiring Oxidoreductases

The most thoroughly investigated structure of a complex NADP
dependent oxidoreductase is that of glutathione reductase (GTHR)
for which crystallographic studies at a resolution of 2.0 Å have
been reported (44). The analysis of GTHR, which in this case
refers to the form with bound FAD, has been supplemented with
studies of complexes formed with substrate, with NADPH and of
forms resembling intermediates in the catalytic cycle (46).

Glutathione reductase from human erythrocytes is a symmetri-
cal dimeric protein each subunit consisting of a polypeptide
chain of 478 amino acids (53). The structure of a single subunit
can be roughly divided into four domains which in structure-
function terms have been called the FAD binding domain, the NADP
binding domain, the central domain and an interface domain, the
latter forming the contact surface between the two complementary
subunits (45).   In terms of the amino sequence, the FAD binding
domain includes residues 19-157, the NADP binding domain residues
158-293, the central domain residues 294-364, and the so-called
subunit interface domain consists of the remainder of the polypeptide chain (44).

Analysis of the two dinucleotide binding domains contained
in GTHR revealed that their structures are folded in a similar
manner (54).   The conformation of the FAD binding domain with its
bound coenzyme is shown in Figure 1F, and the topological diagram
of the chain folding is again included in Figure 2.   Figure 1G
contains a stereoview of the NADP binding domain, again in an
orientation roughly the same as the other enzymes described in

Figures 1A to 1E. Since no coordinates for NADPH bound to GTHR
have been published, the coenzyme has not been included in Figure
1G. As is the case for the other dinucleotide binding proteins
the central core of both the FAD and NADP binding domains is a
parallel β-sheet structure. In the case of the FAD binding
domain, only four parallel strands are found in the core β-sheet.
The intervening structure between these strands differs notably
from those described above for the other simple NAD/NADP dependent
oxidoreductases. In particular, the connection between strands
βC and βG of the core β-sheet consists of a three-stranded
antiparallel β-sheet conformation containing the elements βD, βE
and βF. The plane of this antiparallel sheet, βD-βE-βF, is
roughly parallel to the core β-sheet structure formed by segments,
βC-βB-βA-βG (44). The structural homology between the FAD and
NADP binding domains is readily visible from a comparison of
Figures 1F and 1G (54,55). Note that these two nucleotide binding
domains contain both the core parallel and the intervening antiparallel β-sheet conformations in approximately the same orientation.

    The major difference between the FAD and NADP binding domains
appears in the connection between strands, βB and βC. As can be
seen in the stereo drawing of the FAD binding domain in Figure
1F, this connection is comprised of two long α-helices running
antiparallel to each other. In the NADP binding domain only one
helix forms this connection. Schulz has made a careful comparison
of these two domains and has assessed the probability that such
similarities could have occurred randomly (54). The results
indicated that the RMS distance between corresponding α-carbons
in the FAD and NADP binding domains of GTHR is 2 Å. This value
was calculated after optimizing the orientation of the domains
and allowing for appropriate deletions and insertions of amino
acids to maximize the overlap of secondary structure (54). Using
randomly folded globular polypeptide chains as an estimate of

the probability of chance homology (55), the similarities between
the FAD and NAD binding domains of GTHR have been shown to be the
result of gene duplication (54).

The high resolution structure of another complex NADP
dependent oxidoreductase has also been reported (48). p-Hydroxybenzoate hydroxylase (PHBH) is a FAD dependent monooxygenase that
utilizes NADP as a coenzyme. PHBH is a dimer consisting of two
identical polypeptide chains, each of 394 amino acids (56). In
a manner somewhat similar to GTHR, the structure has been divided
into three domains (48). One of these domains is involved in the
binding of FAD and in particular the ADP moiety of this dinucleotide. The binding mode for NADP is unknown at this moment whereas
a preliminary description has appeared for interaction of the
other dinucleotide FAD (48). The domain that binds this cofactor
contains three β-sheet structures; a larger parallel β-sheet,
sheet A, and two 3-stranded antiparallel β-sheets, sheet C and D
(48). The topological arrangement of sheets A and C show a
resemblance to that described for the two dinucleotide binding
domains of GTHR (Figures 1F and 1G) (44,48). Not only are β-
sheets A and C roughly parallel to each other as is the case in
GTHR but the overall topological arrangement of the β-strands and
of the two helices forming the intervening β-strand connections
in the parallel β-sheet is similar. In PHBH the connection
between strands βB and βC of the parallel β-sheet A contains a
rather long insertion as in both dinucleotide binding domains in
GTHR. As mentioned above the binding mode of NADP to PHBH has
not yet been described. However, inspection of the published
stereo-diagram of the structure does not reveal any additional
structural features that bear much resemblance to any of the
dinucleotide binding domains described previously. Of the two
remaining domains, one contains a five-stranded antiparallel β-
sheet with one α-helix forming one of the strand connections
(48). The third domain is composed almost entirely of α-helices

(48). One, therefore, has to conclude that the binding of NADP
to this enzyme is facilitated in a unique manner and may involve
structural features previously unreported.

Finally, it should be mentioned that the NADP binding domain
of ferredoxin-NADP$^+$ oxidoreductase contains a four stranded
parallel β-sheet structure with a topology identical to that
described for DHFR and GTHR. NADP binds to this domain with the
pyrophosphate moiety located near the carboxyl end of the first
β-strand and the amino end of the first α-helix (50).

ENZYME-COENZYME INTERACTIONS

The structure of the enzymes described in the previous
sections demonstrates a degree of structural homology in binding
domains which seem to consist of a core four-stranded parallel β-
sheet motif with at least one interconnecting α-helix. This
structural homology is also reflected at the level of interactions
between the enzymes and their bound coenzymes. In general, it
appears that the adenine ribose part of the coenzyme displays the
most homology and the nicotinamide ribose or the isoalloxazine
ribitol units the least homology in their respective enzyme
interactions. It should be mentioned at this point that the
nicotinamide ribose moiety and in particular, the nicotinamide
ring is not only in contact with the coenzyme binding domain of
any oxidoreductase but also with portions of a catalytic domain
if present. This is noteworthy since the conformations of the
respective catalytic domains share few similarities amongst this
class of enzymes. The cofactor, FAD, has the adenine-ribose-
pyrophosphate moiety in common with NAD. As exemplified by GTHR,
the FAD-enzyme interactions appear to have some similarity with
those described for the binding of the adenine ribose pyrophosphate moiety of NAD to their respective enzymes.

Both NAD and NADP bind to the core nucleotide binding domain
in an extended conformation (34,42,50). The planes of the two

aromatic rings of the bound coenzyme are approximately 10–15 Å
apart and roughly perpendicular to each other. This conformation
is quite different from that believed to prevail for the free
coenzyme and it is also different from that observed in crystals
of the lithium salt of NAD (57). Furthermore the observed conformations of enzyme bound NAD and NADP may be slightly unfavorable from an energetic point of view (57). Thus the specific
interactions between coenzyme and enzyme accompanying binding
must stabilize a conformation of the coenzyme that is somewhat
unfavorable and hence unlikely to be the predominant form in
solution.

     Considering the degree of allowable torsional freedom, the
similarities in the conformation of NAD when bound to sMDH, LDH,
LADH and GAPDH is striking (34). With one important exception,
the differences in conformational parameters might very well be
attributable to experimental errors inherent in the electron
density maps as well as small differences due to interpretation.
The one exception is GAPDH where the nicotinamide ring assumes
the syn conformation. This point will be discussed in more
detail in a subsequent section.

     In addition to the similarities, the dinucleotide binding
domains in DHFR and sMDH for example, show some rather significant
differences. This fact is reflected in the differences in the
conformation of NAD and NADP when bound to sMDH and DHFR, as
shown in the stereo-diagram given in Figure 3. Both enzymes
transfer a hydride ion to and from the "A-side" of the pyridine
ring. In Figure 3 the adenine ring and its attached ribose are
in roughly the same orientation for both NAD and NADP. Note,
however, that beyond the adenine ring and the adenine ribose, the
conformations of the two coenzymes are quite different (42). The
largest difference occurs about the torsional angle around the
bond linking the O5' oxygen atom of the nicotinamide ribose with
the corresponding phosphorus atom of the pyrophosphate bridge

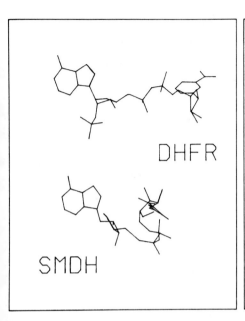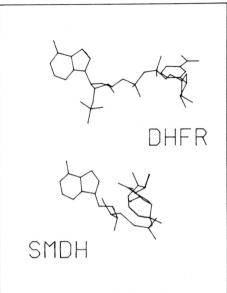

Fig. 3:   Stereo diagrams of NAD as bound to sMDH and NADPH
as bound to DHFR. The two dinucleotides were aligned on the MMS-
X graphics system, with their adenosine moieties in a common
orientation and the adenine ring in the plane of the drawing.

(42).   This is called the psi torsional angle in nucleotide
conformational analysis (58). As can be seen in Figure 3, the
effect is to move the nicotinamide end of the NADPH in DHFR to a
location which is different from that of NAD bound to the 2-hy-
droxy acid dehydrogenases. Nevertheless the separation between
the aromatic rings is about the same. The conformation of FAD
when bound to GTHR also differs substantially from those of
enzyme bound NAD and NADP (compare Figures 1A and 1F). Again the
difference is due mainly to different torsional angles at or near
the pyrophosphate bridge.

     If one ignores the two aromatic rings on both ends of NAD or
NADP, the remaining portion of the coenzyme consists of the

ribose-pyrophosphate-ribose moiety. The latter contains twelve
polar oxygen atoms most of which appear to form hydrogen bonds
with atoms in the oxidoreductase (see for example reference 9).
For sMDH, roughly one-half of these corresponding hydrogen bonding interactions involve acceptor or donor atoms which are part
of the polypeptide chain (9). Unlike hydrogen bonding atoms in
the side chains, those in the main chain are probably less subject
to conformational changes. In terms of the nucleotide binding
domains discussed in the previous sections, this core unit,
independent of the amino acid sequence, provides about one-half
of the non-covalent interactions which occur upon binding of NAD
or NADP to an oxidoreductase. In the sections which follow a more
detailed description of the entire binding site(s) is given.

Adenine Binding Site

     In general, the adenine ring appears to be bound in a
hydrophobic cavity formed primarily by residues in β-strands βB
and βD. One end of this cavity is accessible to the solvent and
the 6-amino group of adenine appears to be in rather unrestricted
contact with the solvent. Although in some instances hydrogen
bonds to the protein are made (LADH, GTHR, DHFR), this feature is
undoubtedly the reason why affinity chromatography utilizing NAD,
NADP, and even AMP derivatives immobilized at the N6 position of
adenine enjoy such a great success in purification of NAD and
NADP dependent oxidoreductases.

     The interactions in this binding pocket are of a rather non-
specific nature and only the hydrophobic character of the amino
acids appear to be conserved amongst the enzymes of known structure. Other hydrophobic compounds have been shown to bind in
this pocket as well and it is to be expected that most inhibitors
of these enzymes generally will bind in the adenine portion of
the coenzyme binding site.

Adenine Ribose Binding Site

The chemical feature that distinguishes NAD from NADP is found at the O2' adenine ribose hydroxyl group which in NADP carries an additional phosphate group. The difference in the interactions the enzyme makes with the O2' hydroxyl and the O2' phosphate groups respectively seems to provide the molecular explanation for coenzyme specificity. In the NAD binding oxidoreductases a carboxyl group from an aspartate residue forms a hydrogen bond with the O2' hydroxyl group. This aspartate is located at the end of the β-strand βB, and has so far been invariant in all structurally known NAD binding enzymes. The presence of an additional phosphate at the O2' position would disrupt this interaction in NAD dependent oxidoreductases due both to steric crowding and to unfavorable electrostatic interactions. In the only system for which NADP binding has been described in detail, DHFR from L. casei (42), the ribose phosphate of NADP interacts with a guanidinium group of an arginine residue, Arg 43. With the exception of the chicken liver enzyme, where it is a lysine, an arginine is invariably observed in this position in all known DHFR's (43). Interestingly this residue is located at the end of the β-strand that corresponds to the N-terminal β-strand in sMDH, βA. Hydrogen bonds from the O2' phosphate are also formed with the side chains of Thr-45 and Gln-65 in DHFR. Residue 45 is either threonine or serine, and residue 65 is a glutamine or a glutamate in the bacterial and vertebrate DHFR's respectively. In GTHR the adenine ribose hydroxyls O2' and O3' in FAD are hydrogen bonded to the two oxygens of the carboxyl group of glutamate 50. This glutamate occupies a position in β-strand βB that structurally and topologically correspond to that of the forementioned aspartate in the NAD binding domains. In the FAD binding domain of PHBH, a glutamate is found in a corresponding position (51,56).

The other oxygens in the adenine ribose are hydrogen bonded either with the protein directly or indirectly through solvent molecules. Part of the adenine ribose makes a number of hydrophobic contacts with the protein. Of most importance are those contacts made with the invariant glycine located at the C-terminal end of β-strand βA (βB in DHFR). Any amino acid with a side chain at this location would disrupt the coenzyme-enzyme interactions which have been described above.

The Pyrophosphate Bridge

The negatively charged pyrophosphate linkage of NAD or NADP is generally located close to the amino terminal end of an α-helix, αB (αC in DHFR). Several hydrogen bonds are formed between the phosphate oxygens and amido groups from the polypeptide chain located at the beginning of this helix and in the turn connecting this helix and the preceding β-strand. It has been suggested that this α-helix has a dipole moment parallel to the helical axis and that this moment facilitates the nonspecific binding of anions near the N-terminus (59,60). The binding of a negatively charged ligand near the amino-terminal end of an α-helix is not unique to dinucleotide binding proteins and has been described in a variety of proteins (59).

In addition to this α-helix, some but not all of these enzymes have basic amino acids, most often arginines, near the negatively charged pyrophosphate moiety of NAD or NADP. Thus Arg 101 in LDH (35) and Arg 44 in DHFR (42) form electrostatic interactions with the pyrophosphate of the bound coenzyme. A compensating positive charge is not consistently found near the pyrophosphate bridge of NAD or NADP in all of the enzymes. This can at least partially be explained by the fact that this negatively charged moiety is also near the surrounding solvent. Solvation as well as the dipole field of the α-helix αB (or αC) might be sufficient to stabilize the binding of the pyrophosphate moiety as has been suggested for GTHR (47).

Finally, at the C-terminus end of β-strand βD in sMDH, LDH, GAPDH and LADH, is found an invariant glycine which is close to the pyrophosphate bridge of bound NAD. In DHFR, a glycine (Gly 99) is found in a structurally equivalent position in β-strand βE and the amido group of this glycine forms a hydrogen bond with a pyrophosphate oxygen. In the two dinucleotide binding domains of GTHR, glycines are also found in equivalent positions, Gly-157 and Gly-290 (44,47).

Nicotinamide Ribose

Much like the adenine ribose, the nicotinamide ribose is located in a shallow surface crevice where it makes hydrogen bonds and nonpolar contacts with the corresponding protein. Of the nonpolar contacts, those made with the glycine located at the end of β-strand βD (βF in DHFR and βG in GTHR)) are notable. This is the same glycine that was observed to be near the pyrophosphate as mentioned above. In the NAD dependent oxidoreductases, the carbonyl group of the residue preceding this particular glycine is hydrogen bonded to the O3' of the nicotinamide ribose. In DHFR the O3' hydroxyl group is hydrogen bonded to the carbonyl oxygen belonging to His 18. Replacing this glycine with any other amino acid would cause steric interference with the nicotinamide ribose in its known binding position. The O2' hydroxyl group of the nicotinamide ribose is hydrogen bonded to a backbone amido group located at the end β-strand βE in sMDH, LDH and LADH, to a solvent ion in GAPDH and to carbonyl oxygen of His 18 in DHFR. This hydroxyl group also appears to be accessible to solvent.

Nicotinamide Binding Site

The nicotinamide ring is bound in a cavity created by residues in both the coenzyme binding domain as well as in the rest of the enzyme molecule. Of the enzymes which have been described all are A-side stereospecific in their hydride transfer reaction

with the exception of GTHR and GAPDH which have B-side stereospecificity. In all of the enzymes, the side of the pyridine
ring opposite to that for which hydride transfer occurs makes
numerous hydrophobic contacts with the protein. Some but not all
of the residues involved may be structurally equivalent. The
hydride transfer side of the nicotinamide ring faces the substrate,
or in the absence of substrate may be partially accessible to
solvent. In the complex dinucleotide dependent oxidoreductases,
FAD is the substrate for the initial hydride transfer to and from
the nicotinamide ring. In GTHR, the isoalloxazine ring is bound
to the enzyme in a manner such that it is in close contact with
the nicotinamide moiety of the NADP and both rings are somewhat
isolated from solvent. Thus in both the simple and complex
oxidoreductases, the productive complex formed by enzyme, nicotinamide dinucleotide and substrate (or FAD), the nicotinamide
ring and substrate are located near to each other in a micro-
environment which is largely isolated from solvent.

      In all of the enzymes analyzed, the carboxamide group of the
nicotinamide moiety forms hydrogen bonds with the enzyme. These
hydrogen bonds may be thought of as directing the orientation of
the nicotinamide ring into either the syn or anti conformation,
as may be required for stereospecific hydride transfer. However,
the stereospecificity is not simply related to the syn vs anti
conformation of the nicotinamide ring. In sMDH, LDH, LADH and
DHFR, the nicotinamide ring is in the anti orientation, and these
enzymes are all A-side specific. GAPDH and GTHR are both B-side
specific, but the orientation contained in the glycosidic bond
differ for the two enzymes; it is syn in GAPDH (19,21) and anti
in GTHR (46). Chemical modification of the carboxamide group of
NAD or NADP can produce some dramatic changes in the binding of
the coenzyme. This is exemplified by the binding studies of the
pyridine and 3-iodo-pyridine analogues of NAD to LADH (61). In
both instances the pyridine ring is located far from the normal
nicotinamide binding site. That the iodo derivative binds in a

non-productive mode, can be explained by the steric bulkiness of
the iodine atom. However, for the pyridine derivative the most
reasonable explanation for improper binding is the inability to
form the normal carboxamide hydrogen bonds.

In many of the crystalline complexes of nicotinamide adenine
dinucleotide and enzyme, the electron density for the nicotinamide
ring has frequently been of a poorer quality than the rest of the
bound coenzyme (sMDH, LDH, GAPDH) leading to slight uncertainty
about the ring orientation.   On the other hand, the structural
studies of DFHR have been performed at a higher resolution and
hence provide the most definitive results about the structure of
an oxidoreductase with bound nicotinamide adenine dinucleotide.
Within the limits of the resolution, the reduced pyridine ring
when bound to DHFR is planar. The carboxamide group is coplanar
with the six-membrane ring but is rotated by 180° from its most
stable position normally found in solution (42). This conformation
has the amide group in a trans position to the C4 carbon of the
nicotinamide ring and is stabilized through hydrogen bonds between
the amide of coenzyme and the carbonyls of Ala-6 and Ile-13, and
between the carbonyl oxygen of the coenzyme and the amido nitrogen
of Ala-6. Furthermore, the pyridine ring hydrogens at positions
2, 4, 5 and 6 all seem to be closer than normal Van der Waal
contact distances to oxygen atoms in DHFR (42).   Thus the C2 is
3.2 Å away from the carbonyl of Ile-13, C6-carbon is 3.3 Å away
from hydroxyl of Thr-45 and the C4 and C5 carbons each 3.3 Å
away from the carbonyl of Ala-97 (42). All three protein oxygens
are approximately in the plane of the nicotinamide ring. These
unusually short contact distances suggest that the pyridine ring
atoms of the enzyme bound coenzyme in DHFR may be oriented by
C-H ----- O types of hydrogen bonds (42) although such noncovalent
bonds have not been observed in the other dinucleotide dependent
oxidoreductases.

In the GTHR-NADPH complex (46) the reduced nicotinamide ring
appears planar with the carboxamide group being twisted about 20°

out of the ring plane. The nicotinamide ring is nearly parallel
with and in Van der Waals contact with the flavin ring of FAD.
The amide group of the carboxamide is hydrogen bonded to the
carbonyl oxygen of Val-370 and OE2 of Glu-201. The carbonyl
oxygen of the carboxamide is not reported to be involved in any
hydrogen bonding but is positioned trans to the C4 carbon of the
nicotinamide ring. This orientation is opposite to that observed
in DHFR. The C4 carbon of the nicotinamide ring is close to N5
of the flavin ring (3.5 Å away), the amino group of Lys-66 (2.6
Å) and OE1 in the carboxyl group of Glu-201 (2.3 Å) (46). The
last distance is remarkably short.

    In the NAD-dependent dehydrogenases, the carboxamide side
chain of the nicotinamide ring is hydrogen bonded to backbone as
well as side chain atoms. In sMDH the carboxamide is hydrogen
bonded to the carbonyl oxygen of residue 132 and the side chain of
residue 149 which is thought to be a serine (9,39), in LDH to the
carbonyl oxygen of Ser-139 and the side chain of Ser-163 (35), in
LADH to the backbone amido group of 319 (26) and in GAPDH to the
side chain of Asn-313 (19). When the crystallographic refinement
of these structures has progressed further it should be possible
to evaluate these interactions with more certainty.

THE CONFORMATIONAL RELATIONSHIPS AMONG THE DINUCLEOTIDE BINDING DOMAINS

    As already noted, one of the most significant observations
that emerged from the early studies of the NAD dependent enzymes
was that sMDH, LDH, GAPDH and LADH all had coenzyme binding
domains, that were structurally homologous. With the completion
of crystallographic studies of other dinucleotide dependent
oxidoreductases, it became clear that the majority of enzymes
that bind dinucleotides and for that matter mononucleotides can
frequently be divided into two or more domains. Numerous attempts
have been made to determine common protein conformations responsible for the binding of the dinucleotides NAD, NADP and FAD as

well as mononucleotides such as AMP, ADP, ATP, FMN, etc. to their
relevant enzymes. Even though some patterns of structurally
similarity appeared to exist, the comparisons have resulted in a
great deal of confusion regarding evolutionary homologies. To a
large degree, this confusion arose because these coenzyme binding
domains contained similar elements of super-secondary structure
which may be the result of energetic rather than evolutionary
factors (62). Even at this point in time, the structural com-
monality between nucleotide binding sites in different proteins
remains obscure.

In most such proteins the nucleotide binding site is located
in a domain that is centered around a parallel β-sheet structure,
but in some instances the corresponding sheet includes β-strands
in an antiparallel orientation. The number of strands in the core
sheet range upwards from three depending on the enzyme. The
connections between the strands are with rare exceptions all of
the right-handed type and are frequently, but not always, alpha
helices. If, for the moment, the topology and the type of connections between the elements of secondary structure is ignored,
a structural motif consisting of a β-sheet flanked by predominantly α-helices is commonly observed in many proteins, some of
which do not bind nucleotides. In order for a structural motif
in one protein to be related to that in another at both the
functional as well as at the evolutionary level, several criteria
must be fulfilled. For instance, the elements of the secondary
structure must appear in the same relative order along the polypeptide chain and their topological distribution must be the
same. That is, the type and handedness of connectivity must be
the same. It is equally important that structurally equivalent,
or nearly equivalent residues perform the same, if not identical,
functional roles.

The original formulation of the NAD binding domain was based
on the existence of a six-stranded parallel β-sheet. In a simplified manner this structure could be described as being composed

of two β-α-β-X-β units. In many instances X̲ would be another α-
helix, thus forming the so-called "Rossmann fold". The NAD
binding domains of sMDH and LDH are examples of this motif. As
discussed in a previous section, and illustrated in Figure 2,
LADH and GAPDH contain basically the same motif, although some
minor deviations are observed, at the left and right edges of the
β-sheet. Part of this structural motif is also found in the NADP
requiring enzymes. Thus DHFR and both dinucleotide binding
domains in GTHR contain the β-α-β-X-β conformation. In this
"pseudo Rossmann fold" X is one or two helices in the two dinucleotide binding domains of GTHR and an irregular polypeptide
conformation in DHFR. However, the core β-sheet domain is in all
cases augumented with an additional β-strand running parallel and
adjacent to the β-strand which starts the core sheet structure.
Hence, the conformation common to all of the dinucleotide dependent oxidoreductases is a four-stranded parallel β-sheet domain.
The connectivity of this folding unit using the nomenclature of
Richardson (33) can be described as +1, +1, -3.

Returning to the simple NAD binding oxidoreductases, inspection of Figure 2 shows that an identical topological folding unit
is found in sMDH, LDH, LADH and GAPDH, but that it is a subset of
the 6-stranded core domain which was originally proposed. The
basic feature of this proposed dinucleotide binding domain thus
consists of a four-stranded parallel β-sheet with an α-helix
forming the connection between the first and second β-strand.
Using the nomenclature applied to sMDH (Fig. 2) the elements in
this motif are βA, βB, βC and βD with helix αB connecting the
first two β-strands. The suggested common structural elements in
these dinucleotide binding domains are indicated in Figures 1A-G
with a heavy outline of the backbone and by the stipling in
Figure 2. As can be most easily seen in Figure 2, the topological distribution of the four structural elements in this proposed

dinucleotide binding domain is the same in all structurally known NAD and or NADP dependent oxidoreductases.

The amino acid sequences of these five units of secondary structure are listed in Figure 4. The alignment, which is an extension of those previously reported for the NAD binding proteins (4,30,34) and of mMDH and sMDH with LDH (39), shows the full sequence from the beginning of strand βA to the end of βB, but only the segments containing βC and βD. The residue number of the first and last amino acid in each segment is also indicated. The alignment of the amino acid sequences of mMDH, LIPDH and PHBH are based on amino acid homologies and not optimized conformational similarities (39,51). The principal criteria employed in this alignment was that homologous side chains in the β-sheet all point to the same side of the β-sheet. Similarly the single α-helix included here was aligned such that homologous side chains were pointing in the same relative direction (34). A preliminary alignment was made by inspection of the molecular models using a MMS-X graphics system and further checked by least squares methods, following the procedures previously described (36). The two nucleotide binding domains of GTHR align quite well with those of the simple NAD dependent dehydrogenases. Thus when the coordinates for the 30 α-carbon atoms in the core sheet used in the amino acid sequence alignment are compared a mean deviation of 1.6 Å is obtained when LDH is compared with the FAD binding domain of GTHR, and 2.1 Å when the helix is included. When comparing the core structures in the NADP binding domain of GTHR and LDH the corresponding numbers are 1.6 Å and 2.0 Å. When the two domains of GTHR are compared with each other the mean deviations are 1.0 Å and 1.2 Å respectively. The largest discrepancies are observed at the ends of the secondary structural elements. For example, the length of a given β-strand may differ by one or two residues from one protein to another. The poorest

```
                                 *    *   *                               *
mMDH        1    A K V A V L G - A S G G I G Q P L S L L L K N S P L V S R L T L Y D   33  V K G Y L G P   59  C D V V V I P A G   77  *
sMDH        5    I R V L V T G A A Q L A F T L L Y S I G B G S (-----) I L L S L M D   40  L K S Z F G K   62  Z B V G V L L A G   85
LDH         22   N K I T V V G V G Q V G M A C A I S I L G K S - - L T D E L A L V D   53  P K I V A N K   82  S K I V V V T A G   99
LADH        193  S T C A V F G L G G V G L S V I M G C K A A G - - A A R I I G V D       223  G A T E C V N ·  236  V D F S F E V I G  270
GAPDH       1    S K I G I N G F G R I G R L V L R A A L S C G - - A Q V V A V N D       32   K K I T V F N    69  A E Y I V E S T C   97
CTHR(1)*    21   Y D Y L V I G G G S C G L A S A R R A A E L G - - - A R A A V V E       50   S H I E I I R   121  A P H I L I A T G  157
CTHR(2)     188  G R S V I V G A G Y I A V E M A G I L S A L G - - S K T S L M I         217  A G V E V L K   241  V D C L L W A I G  290
LIPDH*           A D V T V I G S G P G G Y V A A I K A A Q L G - - - F K T V C I E
PHBH*       3    T Q V A I I G A G P S G L L L G Q L L H K A G - - I D N V I L E         32
DHFR        37   K I M V V G R R T Y E S F P K R P L P E - - - - - R T N V V L T         63   Q G A V V V H    71  E L V I A C G A Q  101

                 |_____|                |_____|      |_____|        |_____|
                               βA              αB                         βB                   βC                    βD
```

Fig. 4: Amino acid sequences of the core dinucleotide binding domains. The sequence number of the first and last residue included in each element of secondary structure are located above each sequence. The secondary structure assignments are indicated by the solid bars at the bottom of the figure. The exact starting points for the β-strands and the end point for the helix differ somewhat from enzyme to enzyme. The stars, (*), at the top of the figure indicate the location of conserved amino acid residues as described in Tables 1 and 2. Abbreviations for enzyme names are those used in the text with the following additions: mMDH, mitochondrial malate dehydrogenase; GTHR(1), FAD binding domain of glutathione reductase; GTHR(2), NADP binding domain of glutathione reductase; LIPDH, lipoamide dehydrogenase; * after an enzyme name indicates a FAD binding domain. The symbol, (----), indicates the position of an insertion of 7 amino acids, (VFGKBZP).

fit was obtained when comparing the DHFR structures with the
other proteins.   In particular the helix component of the core
structure is in a slightly different orientation relative to the
β-sheet.   The mean derivation is 2.4 Å when comparing the core β-
sheets in DHFR and LDH and 3.5 Å when the α-helix is included.

The structural homology including the connectivity of the
core four-stranded sheet structure seems firmly established and
becomes even more compelling when the coenzyme-enzyme interactions
are considered.   Equivalent secondary structural elements amongst
the different oxidoreductases appear to make similar if not
identical interactions with equivalent parts of the bound dinucleotide, whether it is NAD, NADP or FAD.   This is shown in a schematic manner in Figure 5.

One of the common features of these binding sites appears to
be a β-strand, βA, followed by a turn and a segment of α-helix αB
which can be represented as β-T-α.   Except for DHFR the amino
acid sequence segment G-X-(G,A)-X-X-G is found in this turn.   The
pyrophosphate moiety of the dinucleotide (or nucleotide) can come
quite close to the polypeptide chain especially near the middle
glycine or alanine position in the interconnecting turn.   Any
other amino acid here would probably prevent this from taking
place.   Present in the turn and at the beginning of the α-helix
are a number of free amide and carbonyl groups which can form
hydrogen bonds with the pyrophosphate and ribose oxygen atoms of
the bound coenzyme.   Depending on the number of residues in the
turn, $\underline{m}$, there are $2m + 3$ such hydrogen bonding groups or $m + 3$
hydrogen donors.   The factor of 3 comes from the unpaired amide
hydrogens free on the N-terminal end of the α-helix member. As
already noted, the α-helix of this super-secondary structure does
not always run in a direction parallel to the β-strand, but
nonetheless can provide hydrogen bonding atoms for interacting
with the dinucleotide.

The availability of an excess of hydrogen bond donors and
acceptors in this type of super-secondary structure, allows for a

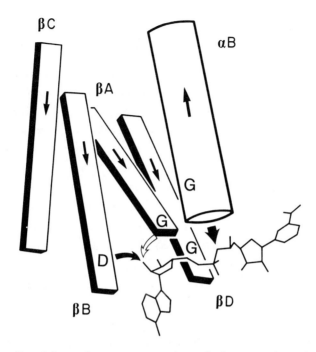

Fig. 5: Schematic representation of the core dinucleotide binding structure. The drawing is based on the sMDH structure and shows the four β-strands as bars, and the single helix represented by the cylinder, together with a stick model of NAD in its location at the active site. The schematic drawing represents the heavier outlined parts of Figure 1A. The direction of the poly peptide chain from N- to C-terminal is indicated by the arrows on the elements of secondary structure. The four amino acid residues that are nearly invariant are shown by the three "G"'s and the single "D" and are discussed in the text. The arrow from the "D" on β-strand, βB, indicates the interaction between the aspartate or glutamate at this position with the O2'-hydroxyl group of the adenine ribose in NAD or FAD. The arrow at the end of the β-strand, βA, indicates the interaction between an arginine (Arg 43) in DHFR and the 2'-phosphate of NADPH. The arrow shown at the end of the cylinder represents the - to + direction of the dipole thought to be associated with the α-helix and which may interact with the negative electrostatic charge associated with the pyrophosphate moiety of the bound dinucleotide.

great deal of "wobble" in the interaction of the ribose-pyro-
phosphate moiety with the relevant segment of polypeptide chain.
Such a wobble is observed with no obvious precise correlation
between the hydrogen bonds formed and the location of the member
atoms in the β-T-α structure. The detailed conformation of the
polypeptide chain in this turn differs somewhat amongst the different dinucleotide dependent oxidoreductases and this may also
contribute to the wobble. Although the number of examples of NAD
or NADP binding domains is still rather limited, it appears that
at least one of three free amide hydrogens on the N-terminal end
of the α-helix forms a hydrogen bond with the pyrophosphate
moiety of the bound coenzyme. As already mentioned, some workers
have suggested that this helix has, in addition, a dipole moment
of sufficient magnitude to serve as a partial counterion for the
negative charge associated with the pyrophosphate moiety (59).

     Looking at Figure 5, three glycines are highly conserved in
the core dinucleotide binding domain. The first glycine is close
to the adenosine moiety of the bound coenzyme, in particular, the
adenine ribose. Since any other residue would affect the positioning of the coenzyme it is not surprising that a glycine is
always found here. In DHFR a glycine is found in a similar
position except it is offset by one residue in the alignment
given in Figure 4. The third glycine in this hexapeptide sequence,
G-X-(G,A)-X-X-G, is not in direct contact with the coenzyme.
Rather it belongs to the helix αB but is close to the core β-
sheet. Any large side chain here would prevent the sharp turn to
be made between βA and αB. It is highly conserved although it
is an alanine in the NADP binding domain in GTHR, and a serine in
DHFR.

     A second common structural feature of the dinucleotide
binding domains is the nearly invariant presence of an amino acid
side chain containing a carboxyl group at the end of β-strand,
βB. This amino acid is indicated with a D in Figure 5 and is
generally found to be hydrogen bonded to the O2'-hydroxyl of the

adenine ribose. It is an aspartate when the bound cofactor is
NAD and a glutamate when bound to FAD. In the NADP binding domains, this residue is an isoleucine in GTHR and a threonine or
serine in DHFR. In the later enzyme, the nonpolar part of the
threonine side chain is in contact with the adenine ring just as
the aspartate side chain is in the NAD binding oxidoreductases.

The glycine located at the end of β-strand, βD, is a third
common but not invariant feature in the dinucleotide binding
domain. This residue, which also is indicated in Figure 5, is
located close to the nicotinamide ribose, and once again a bulky
side chain would affect the positioning of this part of the
coenzyme. Despite the fact that a ribitol rather than a ribose
moiety must be accommodated in the FAD binding domain, both of
the dinucleotide binding domains of GTHR have glycines in homologous positions. Although out of alignment by two residues, a
glycine is found in a roughly equivalent position in DHFR as well
where it is again close to the nicotinamide ribose.

SUMMARY

A large number of proteins belonging to the class of enzymes
called oxidoreductases catalyze the transfer of a hydride ion to
or from nicotinamide and flavin adenine dinucleotides. Since
crystallographic studies have led to models of NAD, NADP and FAD
dependent oxidoreductases, a comparison of this wide spectrum of
enzymes shows that a relatively small segment of each protein has
a common super-secondary structure. This core segment includes
a β-sheet with four parallel strands and at least one α-helical
segment. The connectivity of these elements in terms of the amino
acid sequence is the same for all of the proteins and hence also
compatible with an evolutionary relationship.

In terms of coenzyme binding, the interaction of the adenine
ribose end of NAD, NADP or FAD, occurs close to the same β-strand
in the core domain. This β-strand is the one connected to the
reoccurring α-helical segment of the core structure. Hydrogen

bonding atoms from the polypeptide chain itself which are located
at the N-terminal end of the helix plus similar atoms in the turn
account for part of the stabilizing energy in the binding process.
Amongst the proteins having this common core domain, there seem
to be variations in the precise conformation of the polypeptide
chain in the turn and at the beginning of the α-helix. This
appears to lead to small variations or a "wobble" in the hydrogen
bonding pattern between the dinucleotide and each specific enzyme.

The core conformation requires the positioning of small side
chains, either glycine or alanine at definable locations. This
is necessary to permit the dinucleotide to be close enough to the
core structure to permit the forementioned pattern of hydrogen
bonds, some of which include polypeptide chain atoms. In addition the NAD dependent oxidoreductases all contain an acidic
residue near the O2' hydroxyl group of the adenine ribose as do
those that bind FAD. A basic amino acid replaces this residue in
the NADP dependent oxidoreductases. However, its location in
the core structure is different.

Syn or anti conformations of the glycosidic bond linking the
ribose with the nicotinamide ring account for the differences
observed in the stereospecific transfer of the hydride ion to and
from the nictoinamide ring. However, it appears that the resulting transfer, A-side or B-side, is not correlated with the conformation at the glycosidic link but rather results from the location of the rest of the active site and the position of the
second substrate. Hence the anti conformation does not always
correlate with A-side transfer enzymes and vice versa.

The carboxamide side chain of the nicotinamide ring is
consistently involved in hydrogen bonds with every oxidoreductase
but not with atoms in the core structure. On the other hand, the
adenine moiety of NAD, NADP or FAD appears to be in a hydrophobic
environment largely associated with the core dinucleotide binding
domain. This leads to a paradoxical situation for those interest-
ed in enzyme mechanisms. The portion of each dinucleotide depen-

dent oxidoreductase molecule most involved in the reversible
transfer of a hydride ion to another substrate is also the part
of the protein structures which are most different in this
"family" of enzymes.

Acknowledgements

This work was supported by National Science Foundation Grant
PCM-7921864, and a United States Public Health Service Grant GM-
13925. The MMS-X system was in part supported by United States
Public Health Service National Institutes of Health Grant RR-
00396. The excellent assistance by Sophie Silverman and Suzanne
Winkler in the preparation of this manuscript is gratefully
acknowledged.

REFERENCES

1. Dalziel, K., "Kinetics and Mechanisms of Nicotinamide-Nucleo-
    tide Bound Dehydrogenases", in The Enzymes, 3rd Edition, Vol.
    11A, P.D. Boyer, Ed., pp. 1, Academic Press, New York, 1975.

2. You, K., Arnold, L.J., Allison, W.S. and Kaplan, N.O., "Enzyme
    Stereospecificities for Nicotinamide Nucleotides", Trends
    Biochem. Sci., 3, 265 (1978).

3. You, K.-S., "Stereospecificities of the Pyridine Nucleotide-
    Linked Enzymes", Meth. Enzym., 87, 101 (1982).

4. Eventoff, W. and Rossmann, M.G., "The Evolution of Dehydroge-
    nases and Kinases", Crit. Rev. Biochem., 3, 111 (1975).

5. Fink, A.L. and Petsko, G.A., "X-ray Cryoenzymology", Adv.
    Enzymol., 52, 177 (1981).

6. Weininger, M., Birktoft, J.J. and Banaszak, L.J., "Conforma-
    tional Changes and Non-Equivalence in the Binding of NAD$^+$
    to Cytoplasmic Malate Dehydrogenase", in Pyridine Dependent
    Dehydrogenases, H. Sund, Ed., pp. 87, Walter de Gruyter,
    Berlin, 1977.

7. Birktoft, J.J. and Banaszak, L.J., unpublished.

8. Webb, L.E., Hill, E.J. and Banaszak, L.J., "Conformation of
    Nicotinamide Adenine Dinucleotide Bound to Cytoplasmic
    Malate Dehydrogenase", Biochemistry, 12, 5101 (1973).

9. Birktoft, J.J., Fernley, R.T., Bradshaw, R.A. and Banaszak,
    L.J., "The Interaction of NAD/NADH with 2-Hydroxy Acid De-
    hydrogenases", in Molecular Structure and Biological Activity,
    Griffin, J.F. and Duax, W.L., Eds., Elsevier, North-Holland,
    New York, pp. 37, 1982.

10. Adams, M.J., Ford, G.C., Koekoek, R., Lentz, P.J., Jr.,
    McPherson, A., Jr., Rossmann, M.G., Smiley, I.E., Schevitz,
    R.W. and Wonacott, A.J., "Structure of Lactate Dehydrogenase
    at 2.8 Å Resolution," Nature (London), 227, 1098 (1970).

11. Chandrasekhar, K., McPherson, A., Jr., Adams, M.J. and
    Rossmann, M.G., "Conformation of Coenzyme Fragments when
    bound to Lactate Dehydrogenase," J. Mol. Biol., 76, 503
    (1973).

12. Adams, M.J., Liljas, A. and Rossmann, M.G., "Functional
    Anion Binding Sites in Dogfish M_4 Lactate Dehydrogenase,"
    J. Mol. Biol., 76, 519 (1973).

13. White, J.L., Hackert, M.L., Buehner, M., Adams, M.J., Ford,
    G.C., Lentz, P.J., Jr., Smiley, I.E., Steindel, S.J. and
    Rossmann, M.G., "A Comparison of the Structure of Apo Dogfish
    M_4 Lactate Dehydrogenase and its Ternary Complexes," J. Mol.
    Biol., 102, 759 (1976).

14. Grau, U.M., Trommer, W.E. and Rossmann, M.G., "Structure of
    the Active Ternary Complex of Pig Heart Lactate Dehydrogenase
    with S-lac-NAD$^+$ at 2.7 Å Resolution", J. Mol. Biol., 151,
    289 (1981).

15. Musick, W.D.L. and Rossmann, M.G., "The Structure of Mouse
    Testicular Lactate Dehydrogenase Isoenzyme C_4 at 2.9 Å
    Resolution", J. Biol. Chem., 254, 7611 (1979).

16. Murthy, M.R.N., Garavito, R.M., Johnson, J.E. and Rossmann,
    M.G., "Structure of Lobster Apo-D-Glyceraldehyde-3-Phosphate
    Dehydrogenase at 3.0 Å Resolution", J. Mol. Biol., 138, 859
    (1980).

17. Moras, D., Olsen, K.W., Sabesan, M.N., Buehner, M., Ford,
    G.C. and Rossmann, M.G., "Studies of Asymmetry in the Three-
    Dimensional Structure of Lobster D-Glyceraldehyde-3-Phos-
    phate Dehydrogenase", J. Biol. Chem., 250, 9137 (1975).

18. Olsen, K.W., Garavito, R.M., Sabesan, M.N. and Rossmann,
    M.G., "Anion Binding Sites in the Active Centers of D-
    Glyceraldehyde-3-Phosphate Dehydrogenase," J. Mol. Biol.,
    107, 571 (1976).

19. Olsen, K.W., Garavito, R.M., Sabesan, M.N., and Rossmann, M.G., "Studies on Coenzyme Binding to Glyceraldehyde-3-Phosphate Dehydrogenase," J. Mol. Biol., 107, 577 (1976).

20. Garavito, R.M., Berger, D. and Rossmann, M.G., "Molecular Asymmetry in an Abortive Ternary Complex of Lobster Glyceraldehyde-3-Phosphate Dehydrogenase," Biochemistry, 16, 4393 (1977).

21. Leslie, A.G.W. and Wonacott, A.J., "Coenzyme Binding in Crystals of Glyceraldehyde-3-Phosphate Dehydrogenase", J. Mol. Biol. 165, 375 (1983).

22. Biesecker, G., Harris, J.I., Thierry, J.C., Walker, J.E. and Wonacott, A.J., "Sequence and Structure of D-Glyceraldehyde-3-Phosphate Dehydrogenase from Bacillus stearothermophilus," Nature (London), 266, 328 (1977).

23. Eklund, H., Nordstrom, B., Zeppezauer, E., Soderlund, G., Ohlsson, I., Boiwe, T., Soderberg, B.O., Tapia, O., Branden, C.-I. and Åkeson, Å, "Three-Dimensional Structure of Horse Liver Alcohol Dehydrogenase at 2.4 Å Resolution," J. Mol. Biol., 102, 27 (1976).

24. Abdallah, M.A., Biellmann, J.F., Nordstrom, B. and Branden, C.-I., "The Conformation of Adenosine Diphosphoribose and 8-Bromoadenosine Diphosphoribose when Bound to Liver Alcohol Dehydrogenase," Eur. J. Biochem., 50, 475 (1975).

25. Boiwe, T. and Branden, C.-I., "X-Ray Investigation of the Binding of 1,10-Phenanthroline and Imidazole to Horse-Liver Alcohol Dehydrogenase," Eur. J. Biochem., 77, 173 (1977).

26. Eklund, H., Samama, J.P., Wallen, L., Branden, C.-I., Åkeson, Å and Jones, T.A., "The Structure of a Triclinic Ternary Complex of Horse Liver Alcohol Dehydrogenase at 2.9 Å Resolution," J. Mol. Biol., 146, 561 (1981).

27. Cedergren-Zeppezauer, E., Samana, J.-P. and Eklund, H., "Crystal Structure Determinations of Coenzyme Analogue and Substrate Complexes of Liver Alcohol Dehydrogenase: Binding of 1,4,5,6-Tetrahydronicotinamide Adenine Dinucleotide and trans-4-(N,N-Dimethylamino)cinnamaldehyde to the Enzyme", Biochemistry, 21, 4895 (1982).

28. Eklund, H., Samama, J.-P. and Wallen, L., "Pyrazole Binding in Crystalline Binary and Ternary Complexes with Liver Alcohol Dehydrogenase", Biochemistry, 21, 4858 (1983).

29. The Enzymes, P.D. Boyer, Ed., 3rd Edition, Vol. 11A,
     Academic Press, New York, 1975.

30. Ohlsson, I., Nordstrom, B. and Branden, C.-I., "Structural
     and Functional Similarities within the Coenzyme Binding Do-
     mains of Dehydrogenases", J. Mol. Biol., 89, 333 (1974).

31. Branden, C.-I. and Eklund, H.,"Structure and Mechanism of
     Liver Alcohol Dehydrogenase, Lactate Dehydrogenase and
     Glyceraldehyde-3-Phosphate Dehydrogenase", in Dehydrogenases
     Requiring Nicotinamide Coenzyme, J. Jeffrey, Ed., pp. 40,
     Birkhauser Verlag, Basel, 1980.

32. Sund, H., Ed., Pyridine Dependent Dehydrogenases, Walter de
     Gruyter, Berlin, 1977.

33. Richardson, J., "The Anatomy and Taxonomy of Protein Struc-
     ture", Adv. Prot. Chem., 34, 167 (1981).

34. Rossmann, M.G., Liljas, A., Branden, C.-I. and Banaszak,
     L.J., "Evolution and Structural Relationships among Dehy-
     drogenases", in The Enzymes, P.D. Boyer, Ed., 3rd Edition,
     Vol. 11A, pp. 61, Academic Press, New York, 1975.

35. Holbrook, J.J., Liljas, A., Steindel, S.J. and Rossmann,
     M.G., "Lactate Dehydrogenase", in The Enzymes, P.D. Boyer, Ed.,
     3rd Edition, Vol. 11A, pp. 191, Academic Press, New York,
     1975.

36. Birktoft, J.J. and Banaszak, L.J., "The Presence of a Histi-
     dine-Aspartic Acid Pair in the Active Site of 2-Hydroxyacid
     Dehydrogenases", J. Biol. Chem., 258, 472 (1983).

37. Rao, S.T. and Rossmann, M.G., "Comparison of Super-secondary
     Structures in Proteins", J. Mol. Biol., 76, 241 (1973).

38. Eventoff, W., Rossmann, M.G., Taylor, S.S., Torff, H.-J.,
     Meyer, H., Keil, W. and Kiltz, H.-H., "Structural Adaptations
     of Lactate Dehydrogenase Isozymes", Proc. Natl. Acad. Sci.
     USA, 74, 2677 (1977).

39. Birktoft, J.J., Banaszak, L.J., Bradshaw, R.A. and Fernley,
     R.T., "Amino Acid Sequence Homology among the 2-hydroxy Acid
     Dehydrogenases: Mitochondrial and Cytoplasmic Malate
     Dehydrogenases form a Homologous System with Lactate Dehy-
     drogenase", Proc. Natl. Acad. Sci. USA, 79, 6166 (1982).

40. Janin, J. and Chothia, C., "Packing of α-Helices onto β-
     Pleated Sheets and the Anatomy of α/β Proteins", J. Mol.
     Biol., 143, 95 (1980).

41. Bolin, J.T., Filman, D.J., Matthews, D.A., Hamlin, R.C. and
     Kraut, J., "Crystal Structure of Escherichia coli and
     Lactobacillus casei Dehydrofolate Reductase Refined at
     1.7 Å Resolution, I. General Features and Binding of Metho-
     trexate," J. Biol. Chem., 257, 13650 (1982).

42. Filman, D.J. and Bolin, J.T., Matthews, D.A. and Kraut, J.,
     "Crystal Structures of Escherichia coli and Lactobacillus
     casei Dihydrofolate Reductase Refined at 1.7 Å Resolution,
     II. Environment of Bound NADPH and Implications for Cat-
     alysis," J. Biol. Chem., 257, 13663 (1982).

43. Volz, K.W., Matthews, D.A., Alden, R.A., Freer, S.T.,
     Hansch, C., Kaufman, B.T. and Kraut, J., "Crystal Structure
     of Avian Dihydrofolate Reductase containing Phenyltriazine
     and NADPH," J. Biol. Chem., 257, 2528 (1982).

44. Thieme, R., Pai, E.F., Schirmer, R.H. and Schulz, G.E.,
     "The Three-Dimensional Structure of Glutathione Reductase at
     2 Å Resolution," J. Mol. Biol., 152, 763 (1981).

45. Schulz, G.E., Schirmer, R.H., Sachsenheimer, W. and Pai,
     E.F., "The Structure of the Flavoenzyme Glutathione Re-
     ductase," Nature, 273, 120 (1978).

46. Pai, E.F. and Schulz, G.E., "The Catalytic Mechanism of
     Glutathione Reductase as Derived from X-ray Diffraction
     Analyses of Reaction Intermediates", J. Biol. Chem., 258, 1752
     (1983).

47. Schulz, G.E., Schirmer, R.H. and Pai, E.F., "FAD-binding Site
     of Glutathione Reductase", J. Mol. Biol., 160, 287 (1982).

48. Wierenga, R.K., DeJong, R.J., Kalk, K.H., Hol, W.G.J. and
     Drenth, J., "Crystal structure of p-Hydroxybenzoate Hy-
     droxylase," J. Mol. Biol., 131, 55 (1979).

49. Adams, M.J., Helliwell, J.R. and Bugg, C.E., "Structure of
     6-Phosphogluconate Dehydrogenase from Sheep Liver at 6 Å
     Resolution," J. Mol. Biol., 112, 183 (1977).

50. Sheriff, S. and Herriot, J.R., "Structure of Ferredoxin-
     NADP+ Oxidoreductase and the Location of the NADP Binding
     Site," J. Mol. Biol., 145, 441 (1981).

51. Arscott, L.D., Williams, C.H., Jr. and Schulz, G.E., "Pig
     Heart Lipoamide Dehydrogenase and Glutathione Reductase -
     Homology in Each of the Three Domains," in Flavins and
     Flavoproteins, V. Massey and C.H. Williams, Eds., pp. 44,
     Elsevier North Holland Inc., New York, 1982.

52. Matthews, D.A., Alden, R.A., Bolin, J.T., Filman, D.J.,
     Freer, S.T., Hamlin, R., Hol, W.G.J., Kisliuk, R.L., Pastore,
     E.J., Plante, L.T., Xuong, N.-H. and Kraut, J., "Dihydrofo-
     late Reductase from _Lactobacillus casei_", J. Biol. Chem.,
     253, 6946 (1978).

53. Untucht-Grau, R., Blatterspiel, R., Vrouth-Siegel, R.L.,
     Saleh, M., Schiltz, E., Schirmer, R.H. and Wittman-Liebold,
     B., "The Complete Primary Structure of Glutathione Reductase
     from Human Erythrocytes," in Flavins and Flavoproteins,
     V. Massey and C.H. Williams, Eds., pp. 38, Elsevier
     North Holland, New York, 1982.

54. Schulz, G.E., "Gene Duplication in Glutathione Reductase,"
     J. Mol. Biol., 138, 335 (1980).

55. Schulz, G.E., "Recognition of Phylogenetic Relationships
     from Polypeptide Chain Fold Similarities," J. Mol. Evol.,
     9, 339-349 (1977).

56. Wierenga, R.K., Kalk, K.H., van der Laan, J.M., Drenth, J.,
     Hofsteenge, J., Weijer, W.J., Jekel, P.A., Beintema, J.J.,
     Muller, F. and van Berkel, W.J.H., "The Structure of p-
     Hydroxybenzoate Hydroxylase", in Flavins and Flavoproteins,
     V. Massey and C.H. Williams, Eds., pp. 11, Elsevier North-
     Holland, Inc., New York, 1982.

57. Reddy, B.S., Saenger, W., Muhlegger, K. and Weiman, G.
     "Crystal and Molecular Structure of the Lithium Salt of
     Nicotinamide Adenine Dinucleotide Dihydrate", J. Am. Chem.
     Soc., 103, 907 (1981).

58. Arnott, S. and Hukins, D.W.L., "Conservation of Conformation
     in Mono- and Poly-Nucleotides", Nature, 224, 886 (1969).

59. Hol, W.G.J., Halie, L.M. and Sander, C., "Dipoles of the α-
     Helix and β-Sheet: Their Role in Protein Folding", Nature,
     294, 532 (1981).

60. Warwicker, J. and Watson, H.C., "Calculation of the Electric
     Potential in the Active Site Cleft due to α-helix Dipoles",
     J. Mol. Biol., 157, 671 (1982).

61. Samama, J-P., Zeppezauer, E., Biellmann, J-F. and Branden,
     C.-I., "The Crystal Structure of Complexes between Horse Liver
     Alcohol Dehydrogenase and the Coenzyme Analogues 3-Iodo-
     pyridine-adenine Dinucleotide and Pyridine-adenine Dinucleo-
     tide", Eur. J. Biochem., 81, 403 (1977).

62. Banaszak, L.J., Birktoft, J.J. and Barry, C.D., "Protein-
     Protein Interactions and Protein Structures", in Protein-
     Protein Interactions, edited by C. Frieden and L.W. Nichol,
     pp. 32, J. Wiley, New York (1981).

EXPERIMENTAL APPROACHES IN THE STUDY OF
CRYSTALLINE CYTOSOLIC ASPARTATE AMINOTRANSFERASE

C. Craig Hyde, Paul H. Rogers, Patrick D. Briley,*

and Arthur Arnone

Department of Biochemistry,

University of Iowa, Iowa City, Iowa

Carol M. Metzler and David E. Metzler

Department of Biochemistry and Biophysics,

Iowa State University, Ames, Iowa

ABSTRACT

X-ray crystallographic studies of the vitamin B_6-dependent enzyme L-aspartate aminotransferase (AspAT, EC 2.6.1.1) are well underway. The structure of the cytosolic form of the holoenzyme from pig heart is being solved to 2.7A resolution using the method of multiple heavy-atom isomorphous replacement. The electron density maps have been improved by iterative symmetry-averaging and density modification, and construction of the atomic model on an inexpensive computer graphics terminal has been aided by an automated peptide-fitting routine. We are now determining the spatial and dynamic aspects of the transamination reaction's mechanism through studies of several crystalline AspAT substrate and inhibitor complexes. The structures of coenzyme:substrate intermediary complexes occurring in the enzyme's reaction pathway have been obtained by soaking the native AspAT crystals with the substrate analog 2-methylaspartate and the natural substrate L-glutamate. Significant movements of one domain of the AspAT subunit and rearrangements in the position of the coenzyme have been observed in these complexes. Through the combined use of crystallographic data and single-crystal polarized light spectroscopy (linear dichroism) it should be possible to fully characterize the types and amounts of substrate intermediates in our crystals of AspAT.

*Current affiliation: Digital Equipment Corporation, Wheeling, Illinois.

INTRODUCTION

Essential to the metabolism of many nitrogenous compounds is the transfer of amino groups or the reactions known as transaminations. These involve, in effect, the exchange between two carbon skeletons of an aldehyde or ketone group for an amino group. The large class of enzymes responsible for the catalysis of these reactions are the transaminases (or aminotransferases).

Because of its biological significance and availability, mammalian and avian aspartate aminotransferase (AspAT) are the best characterized of the transaminases (1,2). The enzyme is already well understood with respect to its biochemical, physical, and chemical characteristics, and now the structure and activity of AspAT is being analyzed by crystallographic studies currently in progress.

Three crystallographic groups are now independently solving the structures of different crystalline AspAT isozymes. This work represents the first structural elucidation of a pair of cytosolic/mitochondrial isoenzymes. Borisov et al. (3) were first to report a $5\overset{o}{A}$ low resolution electron density map of the chicken heart cytosolic AspAT isozyme and subsequent studies (4,5,6,7,8) have extended the resolution to $3.5\overset{o}{A}$ and $3.2\overset{o}{A}$. Jansonius et al. (9,10,11) are completing a high resolution analysis of the chicken heart mitochondrial form (mAspAT) to $2.8\overset{o}{A}$ resolution. Our group is completing the $2.7\overset{o}{A}$ resolution structure of the pig heart cytosolic form (cAspAT) (12,13,14).

The generally accepted transamination mechanism put forth by Braunstein and Shemyakin (15,16) and independently by Metzler and Snell (17) relies on the presence of the vitamin B_6-derived coenzyme pyridoxal 5'-phosphate. A key feature of the mechanism is that the reaction proceeds through several stable coenzyme: substrate intermediary complexes. One major goal of the crystallographic work is to directly observe as many of these complexes as possible. This should provide a more precise picture of the spatial aspects of the reaction mechanism and may further define catalytic roles for the coenzyme and apoprotein. Since other classes of B_6-dependent enzymes (including decarboxylases, racemases, and those catalyzing beta-elimination/replacement reactions) probably have some mechanistic and structural features in common with the transaminases, the knowledge gained from the crystallographic studies of the latter perhaps can be extrapolated to the other B_6-dependent enzymes.

Of the over 150 macromolecular structures which have been solved to high resolution, AspAT is one of the larger enzymes. We discuss in the following section the methods used to solve this structure, including multiple heavy-atom isomorphous replacement and density modification/phase refinement. Because the enzyme's size makes both the analysis of electron density maps and model building time consuming, an automated approach for fitting an atomic model will be described. This algorithm is part of the molecular graphics program DISPLA developed for operation on an inexpensive graphics terminal (13).

Knowledge of the structure of a macromolecule can lead to a greater understanding of its detailed mechanism of action. While the result of a crystallographic investigation is often regarded as a rather static image of a macromolecule frozen in one particular conformation, x-ray crystallography is being used to study the dynamic aspects of AspAT's mechanism. Many ligand-enzyme complexes studied in both isozymes of AspAT (5,6,11,13,14) show that the binding of ligands to the active site induces movement of an entire domain of one subunit as well as a rotation of the coenzyme. We have also observed and studied the structure of an authentic intermediate in the reaction pathway by soaking of AspAT crystals in solutions of a natural substrate, L-glutamate. In addition, we have been able to trap the "external aldimine" complex using the substrate analog 2-methylaspartate.

Both single crystal spectrophotometry and polarized light spectroscopic (linear dichroism) measurements have been correlated

with the x-ray studies. In this paper we will show how crystallographic and spectral data may be used in a complementary way to
study the structural changes which occur in AspAT due to the
binding of substrates and substrate analogs.

DETERMINATION OF THE STRUCTURE

Multiple Heavy-Atom Isomorphous Replacement

Pig heart cytosolic aspartate aminotransferase crystals are
grown by the vapor diffusion method using a polyethylene glycol
6000 precipitant and a pH 5.4, 40 mM sodium acetate buffer. Seed-
ing techniques are used to achieve very large single crystals
(12). Under these conditions, the enzyme crystallizes in the
orthorhombic space group $P2_12_12_1$ with four asymmetric units per
unitcell. The asymmetric unit is the dimeric enzyme with an
overall molecular weight in excess of 93,000. The two identical
subunits are each comprised of 412 amino acid residues and one
molecule of a vitamin B_6-derived coenzyme (18,19).

The multiple heavy-atom isomorphous replacement (MIR) method
has been widely applied to macromolecular structure determinations
in order to solve the "phase problem" (for a general reference,
see 20). In theory, phases for each reflection can be estimated
by comparing x-ray scattering from the native macromolecule crystal with one isomorphous heavy-atom derivative. In practice,
however, diffraction data are usually collected from as many
derivatives as possible. Statistical analysis of data from several derivatives minimizes random errors to yield the "best"
phases (21).

For an asymmetric unit the size of the AspAT dimer, it is
estimated that two or three bound mercury atoms are needed to
achieve desirable intensity changes of 20% or greater (22). Until
recently, we had been able to prepare only two good heavy-atom

isomorphous derivatives of AspAT through the modification of four
of the five cysteine sulfhydryls in each subunit which were sufficiently exposed to react with the heavy-atom reagents methyl-
mercurichloride (Me-Hg) and p-mercuribenzoate (PMB). Several of
the other heavy-atom reagents tested showed either low reactivity
or the very large intensity changes characteristic of a lack of
isomorphism. Diffraction data from the mercury derivatives were
used to phase an electron density map to a nominal resolution of
2.7Å. Data collection from a third derivative using potassium
dicyanoaurate (I) has also now been partially completed.

Me-Hg reacts with Cys 82 and Cys 191 of each subunit, whereas
PMB reacts with Cys 45 and Cys 82 of each subunit, and also with
Cys 390 from subunit-2. The common binding sites at Cys 82 are
not identical since Me-Hg and PMB bind to the sulfhydryl group
with different orientations of the S-Hg bond.

Even though the native 2.7Å electron density map was phased
using only two derivatives the map quality was surprisingly good.
(A portion of the native map near the enzyme's active site is
shown below in Fig 1 (top).) Except for a few short loops near
the molecule's surface and parts of the NH_2- and COOH-termini, the
course of the polypeptide chain was readily traced through most of
the map. Tracing in poorer regions was greatly aided by referring
to the published preliminary alpha carbon chain tracing for the
chicken heart mitochondrial isozyme (9). The electron density map
was especially clear in the large coenzyme-binding domain and
active center. On the other hand, the image in the smaller flexible domain of subunit-1 was somewhat poorer. As discussed in
detail below, this smaller domain has been observed to undergo a
movement induced by substrate and inhibitor binding. The increased conformational flexibility in this region may account for
its somewhat "fuzzier" appearance in the native electron density
map.

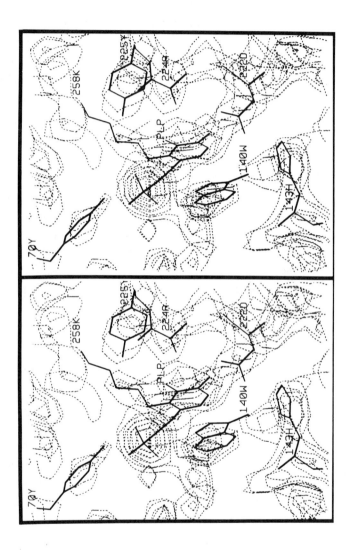

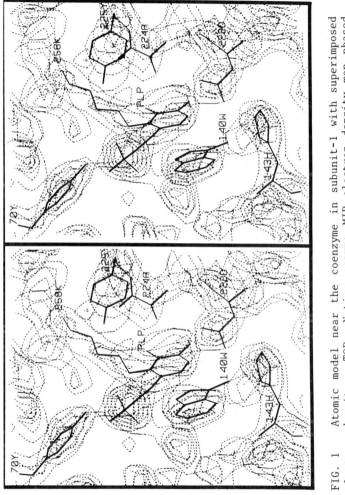

FIG. 1   Atomic model near the coenzyme in subunit-1 with superimposed electron density.   TOP: Native enzyme MIR electron density map phased using two mercury isomorphous derivatives to a nominal resolution of 2.7Å. Diffraction data from a gold derivative to 3.5Å resolution were also included.   Density contour levels are at $+n \cdot \sigma$ ($n=1,2,3,\ldots$), where σ is the RMS density in the map.   BOTTOM: Electron density map computed to the same resolution using improved phases at the end of phase refinement cycle 6.   Density contours are drawn at the same level in the map above.

Electron Density Improvement Through Symmetry Averaging

A clear electron density map allows for easier and more
accurate model building. One way to improve the electron density
clarity for an oligomeric macromolecule is to average the electron
density over structurally identical subunits within the asymmetric
unit. Since AspAT is dimeric with chemically identical subunits
and since, in the orthorhombic cAspAT crystal, the asymmetric unit
is comprised of these two subunits, this averaging procedure can
be used. Because the subunits are not related by a crystal symmetry axis, however, they cannot be rigorously identical. Nevertheless, as in most other examples where multiple copies of a
macromolecule occur in the asymmetric unit, we have found only a
few isolated regions of the AspAT dimer which may deviate from
essentially exact two-fold symmetry. A schematic diagram of the
AspAT subunit and domain structure is shown in Fig. 2.

An initial estimate for the dyad axis was deduced from a
careful analysis of the refined mercury atom positions. Since it
was assumed that the same heavy-atom modification in both subunits
would occur at positions related by noncrystallographic symmetry,
all the heavy-atoms in the unit cell were checked pairwise for
potential alignment across a unique dyad axis. A skewed orthogonal model coordinate system was defined with the estimated dyad
axis as the Y-axis and the Z-axis was aligned approximately along
the longest dimension of the AspAT dimer and orthogonal to the
Y-axis.

A quick examination of a 2.7Å resolution MIR electron density
mini-map sliced perpendicular to the Y-axis showed that the two
subunits had both a high degree of structural similarity and dyad
symmetry. However, a closer inspection proved that corresponding
density features from either subunit were not exactly related by a
simple two-fold rotation about the Y-axis. Many groups of atoms,
especially those in periphery of the molecule, were "off" by as
much as 2Å when rotated onto the other subunit about the dyad

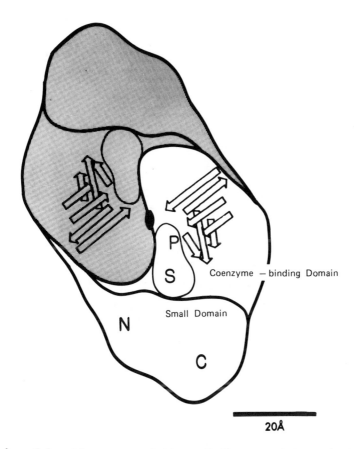

20Å

FIG. 2   Schematic representation of the aspartate aminotransferase domain and subunit structure viewed along the dimer's dyad
symmetry axis.   The two crystallographically-independent subunits
are designated as subunit-1 (unshaded) and subunit-2 (shaded).
The directions of polypeptide strands forming a strongly-twisted
β-sheet at the center of the large coenzyme-binding domains are
shown by arrows.   A pyridoxal 5´-phosphate coenzyme (P) binds to
one side of the central β-sheet.   Each small domain is formed from
the protein's NH_2- (N) and COOH-termini (C).   The substrate-
binding site (S) is formed at the interfaces between large and
small domains and the coenzyme-binding domain of the opposing
subunit.   (cf. α-carbon plot of Fig. 8 shown from this same view.)

axis. Simple symmetry-averaging of the electron density about
this preliminary dyad axis degraded the image.   Since it was not
clear that the AspAT dimer possessed a high degree of two-fold
symmetry, model building of the two subunits proceeded independently.   After a preliminary alpha carbon tracing was completed
and slightly less than half of the 824 residues were model-built,
a better dyad axis was computed.   Coordinates for each subunit's
412 alpha carbons (each estimated to lie within about 1Å of their
true position) were used to compute the dyad axis.   The newer dyad
axis was tilted about 1.6° from the original dyad calculated from
only four pairs of heavy-atom positions.   Small deviations from
two-fold symmetry in one or two of the mercury atom positions were
responsible for errors in the dyad's initial estimate.

Significant sharpening of the image occurred when the electron density map was symmetry averaged about the new dyad axis.
Stereo Fig. 3 (top), showing the symmetry-averaged density, can be
compared to Fig. 1 (top) with nonaveraged density.   The strongest
electron density features were again found in those portions of
the molecule comprising the coenzyme binding domains.   Because
simple symmetry-averaging improved most of the electron density
map, it seemed likely that the bulk of the dimer possessed fairly
rigorous two-fold symmetry.

Phase estimations can often be improved when an asymmetric
unit contains multiple copies of structurally identical, but
crystallographically independent, subunits (23).   Because of the
inherent complexity and computational time involved with this
procedure, we tested further for two-fold symmetry before applying
the density modification and phase refinement method to cAspAT.

We had already observed that about 100 amino acid residues
comprising the flexible domain of subunit-1 can move as much as 3Å
relative to the large domain when substrates and inhibitors are
bound at the active site.   Since many of the weakest portions of
the symmetry-averaged electron density map were also found in this
same region, we suspected that the small domain of subunit-1 may

have undergone a small movement and would not then be symmetri-
cally related about the dyad to its counterpart in subunit-2. Two
different calculations showed that, within experimental error, the
best dyad axes relating either large domains or small domains were
identical to that of the whole dimer.

The first calculation involved a computation of the best dyad
axes through corresponding alpha carbon atoms for both small and
large domains. In addition, a general least-squares transformation matrix was calculated that would rotate and translate the
alpha carbons in either domain from subunit-1 to the corresponding
domain of subunit-2. Despite potentially greater errors in the
small domain's alpha carbon positions, the large and small domain
dyad axes agreed within 0.5°. The general transformation amounted
to a rotation of 180° (± 1.5°) with no net translation.

The second test was made using the program LSQROT provided by
Janet Smith at the Laboratory for the Structure of Matter, U.S.
Naval Research Laboratory (cf. 24). This program refines the
orientation and origin of a rotation axis in an electron density
map by minimizing the sum of the squared deviations between the
electron density at every grid point and the density at its sym-
metrical equivalent. A native electron density map calculated on
a 0.75Å grid was divided into three regions corresponding to solvent, the large domain, and the small domain. The large and small
domains contained over 250,000 and 99,000 grid points, respectively. The best dyad axes for the two domains were refined
independently using the dyad axes derived from the alpha carbon
positions as initial estimates. The procedure converged after 6
refinement cycles for the large domain and after 8 cycles for the
small domain. The orientation of the two dyads agreed within 0.7°
with their origins separated by less than 0.1Å.

A symmetry-averaged vs. native Fourier difference map, where
any deviations from two-fold symmetry would be shown by strong
difference peaks, was used as a final test for dyad symmetry. A
symmetry-averaged, solvent-leveled electron density map, con-

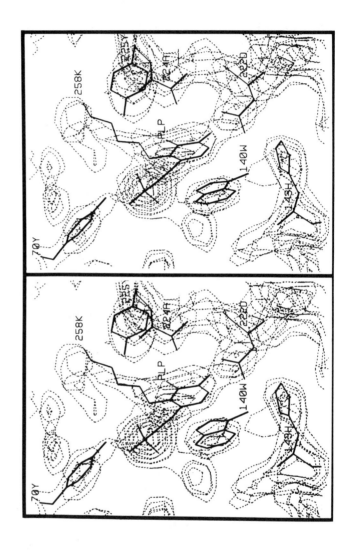

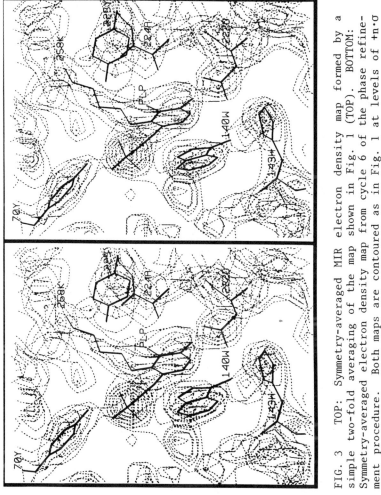

FIG. 3 TOP: Symmetry-averaged MIR electron density map formed by a simple two-fold averaging of the map shown in Fig. 1 (TOP). BOTTOM: Symmetry-averaged electron density map from cycle 6 of the phase refinement procedure. Both maps are contoured as in Fig. 1 at levels of $+n \cdot \sigma$ (n=1,2,3,...).

structed by the density-modification procedure outlined below, was
inverted by Fourier transform to yield calculated "symmetry-
averaged" structure factors scaled against the native observed
structure factors. A difference map calculated using Fobs-Fsym
amplitudes and MIR phases was generally featureless and indicated
that the AspAT dimer does possess a high degree of two-fold symmetry and that the dyad axis had been accurately determined.

    Thus, despite the functional asymmetry observed in substrate
binding (to be discussed below), the bulk of two subunits have
very similar structures in the native, unliganded enzyme. A more
precise analysis of AspAT's non-crystallographic symmetry will
have to await the refinement of the atomic model at high resolution.

Density Modification and Phase Refinement

    There are several examples where investigators have been able
to improve the quality of their electron density maps through
density modification and phase refinement (recent examples include: hemagglutinin glycoprotein of influenza virus (25), beef
liver catalase (26), and aspartate carbamoyltransferase (27)).
Borisov et al. (7) and Harutyunyan et al (6) have also reported
refining phases by variations of this method in their analyses of
the chicken heart cAspAT crystal structures. Initial electron
density maps are modified by imposing such constraints as solvent
constancy, truncation of negative density, and averaging of individual subunits. Structure factors and phases, computed by
Fourier inversion of the modified density map, are statistically
combined with the original phases to obtain new, improved phases.
A new map computed from the improved phases serves as a starting
point for another refinement cycle. The procedure is repeated
until convergence is signalled by only small changes in the phase
angles from the previous cycle.

    The improvement in electron density quality is most dramatic
for crystals either containing a high solvent content or having

many copies of identical subunits (for example, Southern bean mosaic virus (28) and satellite tobacco necrosis virus (29), with 10 and 60 subunits per asymmetric unit, respectively). We have just completed a phase refinement of the $2.7\overset{o}{A}$ resolution AspAT holoenzyme map and have demonstrated that the refinement method can be useful even for AspAT where the solvent content is comparatively low and the asymmetric unit has only dimeric symmetry.

Because the procedure we used was quite similar to that outlined by G. Bricogne (23), the steps are detailed only briefly below:

1) An initial MIR electron density map was constructed on an 0.75A grid using native structure factor amplitudes and phases obtained from the two mercury isomorphous derivatives to $2.7\overset{o}{A}$ nominal resolution. Also included were phases recently derived from a gold derivative to about $3.5\overset{o}{A}$ resolution. The map was placed on an arbitrary scale using $|Fooo|=0$.

2) A molecular boundary, manually traced from both native and symmetry-averaged minimap sections, was entered into the computer with a digitizer tablet. Unit cell grid points lying within the boundary were assigned to protein regions, while all other points were considered to lie in solvent regions. The initial boundary was modified slightly when small overlapping regions with neighboring asymmetric units were detected. Approximately 70% of the over 2 million grid points in the full unit cell map were found within the molecular boundary. Thus, the solvent occupies only slightly more than 30% of the AspAT crystal volume.

3) A new map was constructed with solvent regions set to zero and the protein dimer symmetry-averaged within its molecular boundary about the best dyad axis.

4) The modified density map was inverted by Fourier transform, and calculated structure factor amplitudes were scaled and compared to the measured amplitudes in order to derive computed phase distributions (30). These distributions, combined with the origi-

nal MIR phase distributions, were used to obtain new refined phases and figures of merit.

5) An electron density map, computed from the experimentally measured structure factor amplitudes and new phases, served as the starting point for a new cycle.

Two FORTRAN programs kindly provided by Wayne Hendrickson and Janet Smith of the U.S. Naval Research Laboratory formed the heart of the phase combination procedure. The program EXTRACT was used to express MIR phase probability distributions in compact coefficient form (30,31) while the program COMBIN performed the phase combination of step 4. The dyad axis used for two-fold averaging was refined by the program LSQROT as outlined in the preceding section. Since the dyad's orientation and origin remained essentially unchanged after the first two cycles, the inital dyad position was used throughout. Each cycle required about 1 hour of computer time on a VAX 11/780 computer.

The process converged after six iterations of steps 3 through 5, where the root-mean-square (RMS) phase change between the last two cycles was less than 3°. Fig. 4 summarizes the changes in 18,528 acentric reflection phases after each of the six refinement cycles. Fig. 4 (top) shows the RMS deviation between each cycle's phases and the starting MIR phases as a function of resolution. The overall RMS change in phase angles was about 54°. Fig. 4 (bottom) shows a gradual convergence in the RMS phase changes between successive refinement cycles.

Further statistics for each cycle are given in Table 1. The overall mean figure of merit increased steadily with each cycle. The agreement increased between experimentally recorded structure factor amplitudes and those computed after each cycle as judged by the decreasing reliability factor (R-factor). When plotted as a function of resolution (not shown), the lower resolution structure factors tended to have lower R-factors and higher figures of merit than those in higher resolution classes. The 2,319 centrosym-

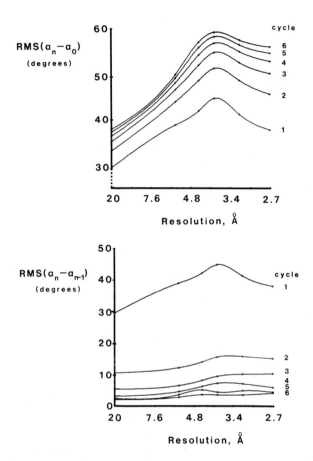

FIG. 4   Density modification/iterative phase refinement statistics.   TOP: Root-mean-square difference between each refinement
cycle's phases, α_n, and initial MIR phases, α_0, as a function of
resolution.   BOTTOM: RMS difference between each cycle's phases,
α_n, and those of the preceeding cycle, α_{n-1}, as a function of
resolution.

TABLE 1

Phase Refinement Statistics

Cycle	Mean FOM*	R-Factor(%)**	2-Fold Difference***
Start	0.692	----	1.41
1	0.907	30.9	0.71
2	0.950	22.8	0.59
3	0.964	19.6	0.52
4	0.969	18.0	0.48
5	0.972	17.0	0.45
6	0.973	16.4	0.42

* Mean figure of merit of 18,528 acentric phases in resolution limits 20Å to 2.7Å. Computed according to Blow and Crick (21).

** The R factor, averaged over all the acentric structure factors, is given by:

$$R = \frac{\Sigma ||Fobs| - |Fcalc||}{\Sigma |Fobs|} \cdot 100\%$$

*** Mean density difference between approximately 350,000 dyad-related grid point pairs lying within the protein molecular boundary, expressed as a multiple of the RMS density of that map.

metric reflections showed generally higher figures of merit and lower R-factors. The decreasing RMS two-fold density differences for each map (shown in the fourth column of Table 1) shows the expected increase in two-fold symmetry with each successive cycle.

The clarity of the electron density map produced after cycle 6 was significantly better than that of the initial MIR map, as shown in Fig. 1. The portion of the molecule shown in the figure is near the coenzyme-binding site and active site of subunit-1. Both the starting MIR map at the top of Fig. 1 and the map computed at the end of cycle 6, shown at bottom of Fig. 1, are contoured at equivalent levels. Since both maps were computed from the same experimentally measured structure factor amplitudes to the same nominal resolution of 2.7Å, the overall increased height of the density features in the lower figure, as well the apparent better resolving power, can be attributed to better estimations of the phase angles.

Several features in the native MIR map shown in Fig. 1 became improved considerably in the map computed with refined phases. In particular, the electron density for the catalytically important residues Lys 258 and Tyr 70 is much better defined. The tight ionic interaction between the coenzyme's pyridinium nitrogen and the carboxylate of Asp 222 is more clearly resolved in the improved map. In general, regions of the map comprising the large coenzyme binding domains showed the greatest improvement. On the other hand, the few nearly uninterpretable peptide stretches in the original MIR map were not significantly better in the final refined map. Since these residues appear to be highly mobile and may deviate from two-fold symmetry, they would not be expected to improve with phase refinement.

As mentioned previously, simple symmetry averaging of the electron density about the best dyad axis also improves the image. Simple symmetry averaging is at least an order of magnitude less time consuming than the iterative phase refinement procedure and also requires no knowledge of the molecular boundary. At first, we were not certain that any significant additional improvement over simple averaging could be achieved by the iterative refinement. Stereo Fig. 3 shows the symmetry-averaged MIR map at top and the symmetry-averaged map of cycle 6 below. The clearer, more intense density features of the map demonstrate that phase refinement improved the map.

A closer inspection of Fig. 4 and Table 1 shows that about one-half of the total change in the mean figure of merit, RMS phase difference, R-factor, and decrease in deviation between two-fold density pairs all occur in the first cycle. If the first cycle is considered equivalent to a simple symmetry-averaging, then roughly half of the total improvement was accomplished by simple symmetry-averaging. Iterations of solvent leveling, symmetry averaging, and phase combination were responsible for further improvements.

Although only a small enhancement of the symmetry averaged native electron density map was observed, the refined phases greatly improved the sharpness of many difference Fourier maps. Because many AspAT substrate and inhibitor complexes are isomorphous with the native PLP holoenzyme, their binding can be readily studied using Fourier difference maps. Since the phases for each reflection from the protein-inhibitor complex and the unliganded protein would not be expected to be very different, difference Fourier maps are computed using the native enzyme MIR phases. Difference density maps recomputed using the improved phases showed higher peaks and generally lower noise. The heavy atom ghost peaks found in difference maps phased with the MIR phases were virtually eliminated in these maps.

Fig. 5 shows a small portion of a difference map computed with both MIR and refined phases. The binding of 2-methylaspartate at subunit-1's active site induces movements in its small domain. Although a difference map using MIR phases at 3.2Å resolution shows many residues in the small domain to move by more than 3Å, the exact nature of the changes is confused by the poor quality of the map. Residues 16 to 23 forming a short helix in the small domain are shown in their position in the native electron density map in Fig. 5 (top). The superimposed difference density, with negative density features around the model and positive features located a few angstroms to the right, does not present a clear picture of the structural changes. In contrast, the density map in Fig. 5 (bottom), computed with refined phases, shows both stronger and better-resolved difference peaks. The more obviously "helical" density features in this map indicate that the helix undergoes a rigid body translation from left to right.

Early difference maps showed that most inhibitors and substrates tested bind preferentially in only one of the two active sites (designated as belonging to subunit-1) and induce a movement of the small domain in that subunit. The amount of inhibitor bound to subunit-1 is usually at least ten-fold higher than that bound to subunit-2 as judged from the difference density peak heights. This same trend was observed in the newer difference maps computed from the refined phases despite our fears that these phases, having been somewhat biased towards two-fold symmetry, would tend to reduce the induced structural differences between the two subunits.

In conclusion, density modification and phase refinement offered added improvement over simple symmetry-averaging in cytosolic aspartate aminotransferase in spite of the low solvent content in the crystal and the dyad symmetry of the asymmetric unit. Even greater improvement may be realized when the procedure is used to extend the current resolution limit. Since the multiple isomorphous replacement method tends to be less reliable in phasing high resolution reflections, the density modification procedure may be a useful way to obtain better phase information as well as to avoid the tedious collection of high resolution isomorphous diffraction data.

Model Building on Inexpensive Computer Graphics Terminals

Conventional model building techniques using wire-frame models and optical comparators have now been largely replaced by modern computer graphic systems. The more sophisticated systems based on real-time graphics are undoubtedly most convenient for fitting an atomic model to electron density maps. Perhaps the only drawback with these systems is their great expense, forcing many investigators to compete for time on shared facilities.

This section describes software that one of us (PDB) has developed to build and display atomic structures on an inexpensive storage graphics terminal (Tektronix model 4006, 1983 list price US $4000). The interactive program DISPLA is written in FORTRAN IV to run on a Digital Equipment Corporation VAX 11/780 computer.

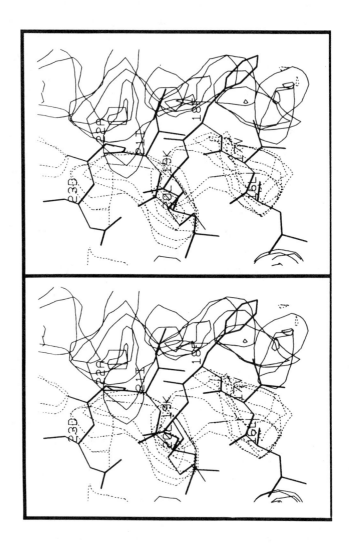

FIG. 5   A portion of the NH$_2$-terminal helix of subunit-1 with superimposed Fourier difference density maps obtained when cAspAT crystals are soaked with 300mM 2-methylaspartate (masp).   TOP: Difference map computed with ‖F masp‖-‖F nativel structure factor amplitudes and MIR phases and figures of merit. BOTTOM: Difference map computed using refined phases. Both maps are phased to the same nominal resolution of 3.2A and contoured at levels of ±n·1.5·σ (n=1,2,3,...).   Positive and negative density features are shown by solid and dashed lines, respectively.

The atomic structures can be built and displayed on any Tektronix-compatible terminal using the PLOT-10 terminal control system. (All of the stereo figures in this paper were drawn using DISPLA.) Some of DISPLA's features include:

-model creation/deletion and manipulation, including rigid body rotation/translation and rotation about bonds

-stereo plotting of model and electron density

-electron density contour plots along any arbitrary direction

-automated peptide fitting into electron density

-line drawings suitable for publication

This molecular graphics system depends to a great degree on the VAX computer. The VAX has a very large virtual address space which allows large programs to be run. The program is approximately 2 megabytes in virtual memory (currently supporting up to 7500 atoms) but can run in very small physical memory (200 kilobytes) without degrading its performance. The user communicates with this program through a uniform command structure with each command entered on a single line. Arguments to each command follow in the form of associated keys. The command language currently consists of 87 different commands and 57 keys.

One useful feature of DISPLA is the way in which electron density maps are displayed. The program draws map contours along sections perpendicular to any desired viewing direction. In this way the map is drawn like the minimap sections familiar to most crystallographers. Both positive and negative difference density can be represented by different line types. We found that the use of multiple contours was needed for obtaining an accurate fit of the coenzyme and important active site residues into both the native and difference density maps.

The basic limitation in this system is its dependence on a storage tube terminal. Any change in the model can be displayed only after erasing the screen and redrawing. While DISPLA has many of the features found in the real-time systems, it only lacks the ability to rapidly display changes made in the model or to rapidly present different views. Superimposing electron density with the model can become especially tedious since the density map is resectioned, contoured, and redrawn with every plot. Because of this limitation, we have tried to automate the most time-consuming part of the fitting procedure.

The BUILD subroutine attempts to fit short peptide stretches into electron density. We have used BUILD to fit approximately 75% of the 824 amino acid residues in the AspAT model. Provided that the electron density map quality is good and that the user can supply adequate starting information, the method works very well. The BUILD routine manipulates the model by rotating about backbone atom phi/psi torsional angles and side chain torsional angles, testing each conformation for a fit to the target positions and electron density. Only short peptide stretches are fit at one time. One feature of this system is that only valid rotation about single bonds are allowed. Consequently, no distortions in the idealized model's bond lengths and angles is ever generated. Since the protein model is constructed in segments, potentially bad geometries created at the junction between peptides can be largely avoided by building in overlapping stretches.

To use BUILD, the user must manually fit the segment's NH_2-terminal residue as accurately as possible, and provide "target" positions for a few atoms further along the peptide. Often only the estimated alpha carbon positions, derived from electron density minimaps, are necessary to help direct the positioning of the peptide. Since the number of possible conformations is very large even for a small peptide, it is impractical to test every one. Rather than conducting a global search of all possible torsional angle combinations, only a small subset of likely conformations is tested as follows. Working in the COOH-terminal direction, the phi/psi pairs are varied in fixed increments (usually by 15° or 20°). In order to further confine the search, the phi/psi pairs are usually restricted to the allowed region of the standard Ramachandran plot (32). For each phi/psi pair, a residual (see below) is calculated using the β-carbon, carbonyl carbon, and carbonyl oxygen of the current residue in addition to the amide nitrogen and α-carbon of the next residue. The phi/psi pairs associated with the lowest N (usually three or four) residuals are saved and serve as the starting points for the conformations which are generated by varing the next phi/psi pair. Thus, successive phi/psi pairs are varied and the lowest N residuals are saved at each step. This process is continued for each of the X amino acid residues in the peptide and N^{X-1} possible conformations are generated. At each of these conformations, the side chain conformations with the lowest residual are found with a similar search by rotating each side chain torsional angle in small increments and selecting the best N side chain conformations at each step. Of the N^{X-1} possible conformations, the "best" conformation is selected as the one with the lowest overall residual.

Even when a only small subset of all the possible conformations is tested in this manner, the number of combinations quickly becomes computationally prohibitive as the peptide length increases. Usually the peptide length is limited to between four and seven residues. If the electron density clearly indicates that a regular secondary structure (β-sheet strand or α-helix) is present, the size of the calculation can be reduced by confining the possible phi/psi pairs to a smaller region of Ramachandran space.

The residual calculated for each conformation is the sum of the three functions listed below:

$$R_1 = \sum_i (\text{target density - density at atom i})^2 / W_1^2 N_1$$

$$R_2 = \sum_i (\text{mean density - density at atom i})^2 / W_2^2 N_2$$

$$R_3 = \sum_i (\text{distance between target and actual position of atom i})^2 / W_3^2 N_3$$

The first function, R_1, is used to keep atoms in strong electron density. The user selects a "target" density level estimated to equal the highest electron density in this region of the map. Residual R_1 serves to minimize the difference between density at all of the atoms and the estimated highest level of the electron density. If an atom should experience a density value higher than the target, the difference is set to zero.

A low value of R_1 can be erroneously obtained when some atoms lie in very strong density, with neighboring atoms placed in weak density. Residual R_2 serves to prevent this by forcing the density to be fairly uniform from one atom to the next. It acts to minimize the difference between the mean density of the residue being treated and the density of each individual atom.

Where possible, information about atom positions can be used to help the build routine prevent the polypeptide chain from "wandering off" into incorrect density. Residual R_3 serves to select conformations that minimize the distance between atom positions and their target positions. An additional "non-bonded contact" residual may be used to prevent collisions between side chains atoms and existing portions of the surrounding model.

The actual overall residual used to select the best conformational angles for the peptide is the sum of the three residuals where each residual's contribution is adjusted by the weighting factors W_1, W_2, and W_3. The factors N_1, N_2, N_3 are the number of atoms that contribute to each summation and normalize each residual to a "per atom" basis. The weighting factors are supplied by the user and usually selected so that the three residuals are given equal weight.

Stereo Fig. 6 shows the results of a typical fitting. Six residues comprising one strand of the AspAT coenzyme-binding domain's central beta sheet were fitted into the 2.7Å resolution symmetry-averaged electron density map. Fig. 6 (top) shows the atomic model both before and after the build. The manually fitted NH_2-terminal residue, Phe 220, and the alpha carbon target posi-

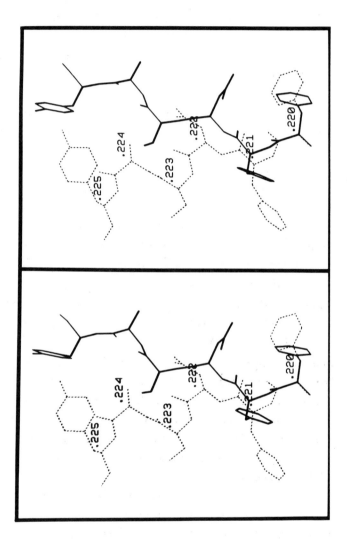

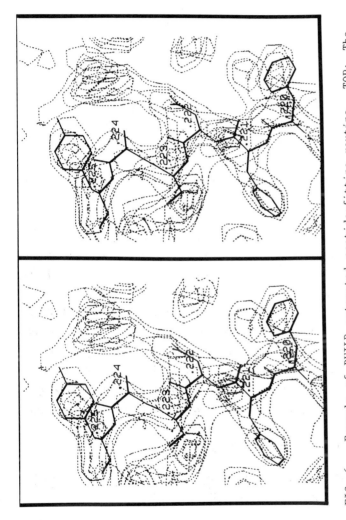

FIG. 6 Example of BUILD automated peptide-fitting routine. TOP: The model's initial extended conformation is shown in dark lines. Only the manually fitted NH$_2$-terminal Phe 220 and the α-carbon target positions for residues 220-225 were provided. The final "best" conformation selected by BUILD is shown in dashed lines. BOTTOM: The symmetry-averaged native electron density map used in the building procedure is shown superimposed on the model's final conformation. (The electron density in the upper right portion of this figure corresponds to the coenzyme ring.)

tions for residues 221-225 were supplied as initial estimates. The model shown in dark lines corresponds to the peptide's extended starting conformation. After two cycles requiring approximately 35 minutes of computer time, the BUILD program selected the conformation shown in dotted vectors as the one with the lowest overall residual. Since the electron density image at first appeared to contain both beta strand and helical conformations, the phi/psi pairs were constrained to the corresponding allowed regions of a Ramachandran plot. These two regions were spanned in 15° increments, resulting in 16 steps in the psi angle and 10 steps in phi. Three conformations with the lowest residuals for each phi/psi pair were selected at each step so that a total of 3^5 main chain conformations were generated for the hexapeptide. Each side chain's torsional angle was tested using 10° increments while retaining the best four conformations at every step. In this particular example, since target positions were provided only for the main chain atoms, side chain fitting relied on only the first two residuals. The weighting factors chosen gave the residuals approximately equal weights.

The final conformation was "fine tuned" in a second BUILD cycle. A set of conformations was generated and tested by rotating each of the final phi/psi pairs by ±15° in small 5° increments. Similar weights and targets were used in the second cycle.

Stereo Fig. 6 (bottom), with the final model superimposed in the electron density, shows that BUILD was able to fold both main chain and side chain atoms into density and prevent the model from diverting from the target positions.

STUDYING ASPARTATE AMINOTRANSFERASE SUBSTRATE AND INHIBITOR COMPLEXES WITH X-RAY CRYSTALLOGRAPHY

The enzymatic transamination reaction proceeds through a complicated, multi-step mechanism with each step being characterized by the formation of a stable intermediary complex. The presence of such complexes makes it possible to directly observe the binding of the true substrates in the crystal. In only a few other examples have investigators succeeded in using crystallographic methods to study true enzyme:substrate complexes (for example: elastase (33), alcohol dehydrogenase (34,35), and carboxypeptidase A (36)). Moreover, because each of the intermediates in the AspAT mechanism are chromophoric, solution and single crystal spectroscopy can be used to identify and quantify the reaction intermediates.

The AspAT reaction and general mechanism is briefly outlined below. A more detailed description can be found elsewhere (1,2).

The reaction catalyzed by AspAT is the sum of two half-reactions:

AspAT/pyridoxal-P   + L-aspartate $\rightleftharpoons$

$\qquad\qquad\qquad$ AspAT/pyridoxamine-P + oxaloacetate

AspAT/pyridoxamine-P + 2-oxoglutarate $\rightleftharpoons$

$\qquad\qquad\qquad$ AspAT/pyridoxal-P   + L-glutamate

Net reaction:

$\qquad$ L-aspartate + 2-oxoglutarate $\rightleftharpoons$ oxaloacetate + L-glutamate

In this reaction scheme the enzyme-bound vitamin B_6-derived coenzyme is interconverted between aldehyde and amino forms, pyridoxal 5´-phosphate and pyridoxamine 5´-phosphate. In the first half-reaction, aspartic acid and pyridoxal 5´-phosphate react to give oxaloacetate and pyridoxamine 5´-phosphate. The second half-reaction involves transfer of pyridoxamine 5´-phosphate's amino group to 2-oxoglutarate, yielding glutamic acid and regenerating pyridoxal 5´-phosphate. The main intermediates occurring in the first half-reaction are shown in Fig. 7 and briefly described below:

Pyridoxal 5´-Phosphate Internal Aldimine: Covalent binding of pyridoxal 5´-phosphate to the apoenzyme yields the catalyti-

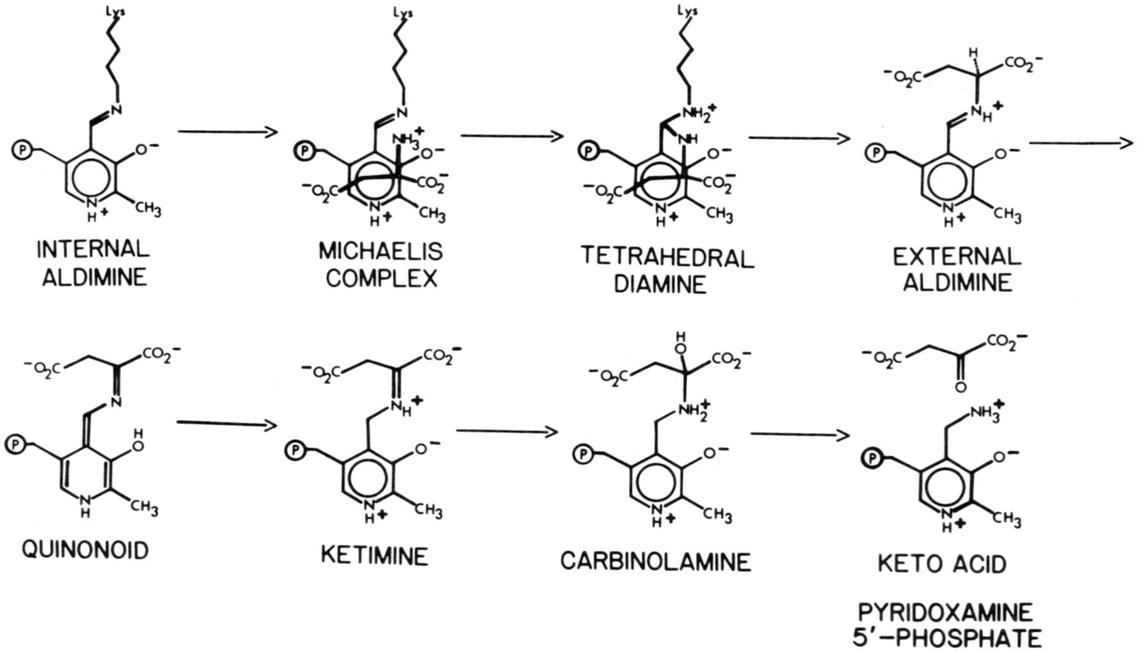

FIG. 7   Structures of most of the stable intermediary complexes formed in the half-transamination reaction converting aspartate to oxaloacetate. When the reaction proceeds as shown, the coenzyme is converted from its aldehyde form, pyridoxal 5′-phosphate (as the internal aldimine) to its amino form, pyridoxamine 5′-phosphate.

cally active holoenzyme. The Schiff base adduct formed between the aldehyde of the coenzyme and the amino group of Lys 258 is referred to as the "internal aldimine" complex. While it is the deprotonated form of the Schiff base that binds the amino acid substrates, the protonated Schiff base predominates in our native enzyme crystals at pH 5.4.

Amino Acid Michaelis Complex: The amino acid substrate is bound so that its amino group is adjacent to the internal aldimine. For reaction to occur, the amino group first donates its proton to the basic aldimine nitrogen.

Tetrahedral Diamine: Nucleophilic attack on the coenzyme's 4´-carbon atom by the substrate's amino group yields a transient geminal diamine adduct. This tetrahedral adduct quickly breaks down by eliminating the Lys 258 side chain to yield a new carbon-nitrogen double bond between the coenzyme and substrate.

External Aldimine: The formation of the new Schiff base completes the "transaldimination" stage, where the "internal aldimine" between coenzyme and enzyme is replaced by an "external aldimine" between coenzyme and substrate.

Quinonoid: The next few steps involve tautomerization of the imine bond. In a partially rate-limiting process (37), the substrate's alpha proton is removed, leaving a carbanion at the substrate alpha carbon atom. The quinonoid intermediate shown represents a stabilized carbanion, where the coenzyme's strongly electron-withdrawing pyridinium nitrogen facilitates resonance delocalization of the carbanion electrons into the conjugated pi-system of the Schiff base and pyridine ring.

Ketimine Intermediate: Reprotonation at the quinonoid 4´-carbon atom yields the ketimine intermediate. The ketimine is a tautomer of the external aldimine where the double bond is now between the substrate's α-carbon and α-amino nitrogen atoms.

Carbinolamine: Addition of a water molecule to the ketimine double bond yields a carbinolamine which can quickly break down to yield the free keto acid product and the amino coenzyme form, pyridoxamine 5´-phosphate (or PMP). This step completes the transamination half-reaction where the substrate's amino group and coenzyme's aldehyde have been, in effect, interchanged.

The overall transamination reaction occurs through a reversal of this half-reaction mechanism when the alternative keto-acid substrate, 2-oxoglutarate, reacts with the PMP form of the co-enzyme to produce the corresponding amino acid, L-glutamate. Since the overall equilibrium constant for the reaction is not far from 1, the reaction is freely reversible.

To date, we have succeeded in collecting diffraction data from crystals soaked with (or grown in the presence of) several different inhibitors and substrate-like molecules. The resulting structural changes, summarized below, show dynamic motion of the protein and coenzyme at different reaction stages. Two of the best-characterized examples, formed by soaking AspAT crystals with 2-methylaspartate and L-glutamate, show the formation of the external aldimine and ketimine intermediates, respectively.

Asymmetry in Induced Domain Movements

Earlier solution experiments have shown that both AspAT sub-units are fully reactive and independent (38,39). In contrast to these findings, we find that only the active site of subunit-1 in the crystalline enzyme is able to bind most substrates and inhibi-tors. Therefore, despite the strong structural symmetry between the two AspAT subunits, this half-sites reactivity clearly shows that functional asymmetry is present in this crystal form of AspAT. Since earlier analyses showed the dimer to possess fairly rigorous dyad symmetry in the crystal, it seems unlikely that inactivity of subunit-2 can be explained by a distortion of that active site. In fact, the best improvements in the electron density maps by symmetry-averaging and phase refinement occurred in the active site region.

We have found, however, that the binding of substrates, most inhibitors, and even acetate anions (at high concentration) at the

active center of subunit-1 induce fairly large changes in the
conformation of a small protein domain formed by the polypeptide's
NH_2- and COOH-termini. We believe this domain movement represents
an essential induced structural change conferring both specificity
and tight binding. Stereo Fig. 8 shows the AspAT dimer's alpha
carbon plot with superimposed 2-methylaspartate difference density. Alternating positive and negative difference density features
show that this domain undergoes a small "right to left" rotation
when the inhibitor is bound at the active site. Although most of
the small domain appears to move as a rigid body, it is difficult
to fully characterize the movements using the low resolution
difference maps. The movement may prove to be more complicated
than a rigid body rotation and involve repacking of several amino
acid side chains. Jansonius et al. (11), reporting a similar
domain shift in their orthorhombic chicken heart mAspAT crystals
in the presence of maleate, have tentatively characterized the
movement as an approximate 12°-14° rotation of the entire small
domain about an axis parallel to the molecular dyad and passing
through Gly 325.

The likeliest explanation for functional asymmetry may be
that weak contacts between surface residues of the small domain of
subunit-2 and those of neighboring subunits in the crystal may
interfere with the domain shift in subunit-2. Because the asymmetric unit is the enzyme dimer, the two subunits maintain different lattice contacts. A search for potential close contacts
between α-carbon atoms in the small domains of either subunit and
those of neighboring molecules showed that subunit-2 maintains far
more potential contacts with neighboring dimers. While only 1
subunit-1 α-carbon was found within 8Å of those in neighboring
subunits, 8 potential close contacts were found in the small
domain of subunit-2.

An alternative explanation might simply argue that access to
the active site of subunit-2 is blocked in the crystal. Visual
inspection of the electron density map indicates that this is not

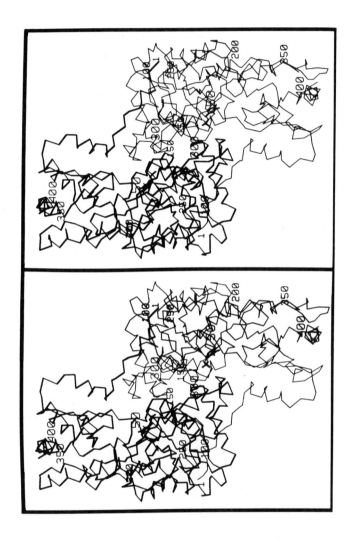

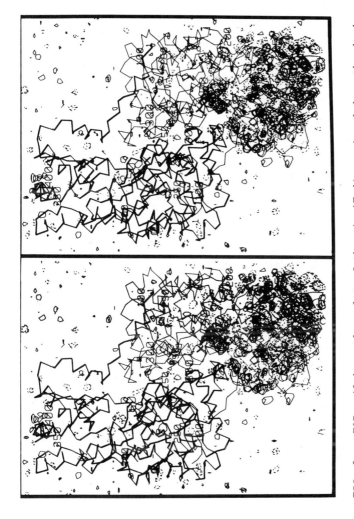

FIG. 8   TOP: Alpha-carbon plot of the cAspAT dimer viewed down the dyad symmetry axis (cf. Fig. 2).   Subunit-2 is drawn in darker lines.   The two PLP coenzymes are shown at the active sites.   BOTTOM: Twelve sections (spaced at 1A intervals) of the 3.2 A resolution methylaspartate vs native difference electron density map superimposed at the level of the small domain and active site.   Contouring is at levels of ±n·2·σ (n=1,2,3,...).

the case. In one experiment, the unreactive inhibitor DL-2-amino-
3-phosphonopropionate at a concentration of 300mM was able to bind
to both active sites but failed to induce movements in either
small domain. While we still do not understand how this compound
prevents the domain movement, the experiment has shown that the
active site of subunit-2 is accessible to ligands which do not
induce domain movement. In at least one other experiment, we
prepared crystal complexes with 2-methylaspartate both by soaking
the inhibitor into a grown native enzyme crystal as well as by
cocrystallizing the inhibitor with the protein! Difference maps
showed that in both cases the binding of the amino acid and the
concomitant domain movement occurred only in subunit-1.

When some substrates or inhibitors are added in high concentration, the crystals crack. Presumably the high concentration
forces the substrate into the active site of chain-2, causing its
small domain to move and disrupt the crystal packing.

Dynamic Motion of Coenzyme and Active Site Residues

Despite the apparent complexity of the transamination mechanism, detailed stereochemical proposals were published well before
crystallographic studies were initiated. A model put forth by
Ivanov and Karpiesky (40) suggested that the coenzyme must rotate
from its position in the native internal aldimine when it forms
the external aldimine complex. The results presented below show
that their proposal was correct in broad outline. Likewise, the
stereoelectronic rules for bond breaking and bond making proposed
by Dunathan (41) are proving consistent with the recent structural
results.

In addition to the widespread conformational changes in the
small domain of subunit-1, the positions of the coenzyme and many
residues forming the active site are altered when substrates and
inhibitors bind and complexes form. The complexes formed are
believed to represent stable intermediates occurring on the main

reaction pathway. Examples of AspAT complexes are presented
below.

2-Methylaspartate: Strong positive difference density present at the active site of subunit-1 shows the binding of this
competitive inhibitor. The complex is formed by soaking native
holoenzyme crystals with 300mM DL-2-methylaspartate at pH 5.4.
Since an earlier study confirmed that holo-AspAT fails to bind
D-2-methylaspartate (42), we presume that only the L-isomer interacts with the enzyme.

Solution studies show that the reaction with 2-methylaspar-
tate passes through the transaldimination stage to yield the
PLP:methylaspartate external aldimine complex. The reaction is
halted at this stage since the natural substrate's alpha proton
(normally removed in the next reaction step) is replaced by a
methyl group.

An atomic model of the L-2-methylaspartate external aldimine,
fit into the 3.2Å resolution difference Fourier map, is shown in
stereo Fig. 9. The map's strongest positive peak, located in the
active site of subunit-1, shows the position of the bound amino
acid analog. Both negative and positive density contours on
different sides of the internal aldimine position indicate that
the ring rotates into the active site to form the external aldimine. The nearly complete absence of difference density around
the coenzyme's phosphate group shows that the phosphate does not
move. Only rotations about the three torsional angles between the
phosphate and aromatic ring were used to reposition the coenzyme
ring.

Although Ivanov and Karpiesky's (40) proposal suggested that
the coenzyme ring may rotate by about 40° about an axis through
C2´-methyl and C5´-methylene carbon atoms, the actual movement is
much more complicated. The angle between the coenzyme ring planes
of internal and external aldimines is about 20°. A small transla-
tional component increases the distance between the pyridinium

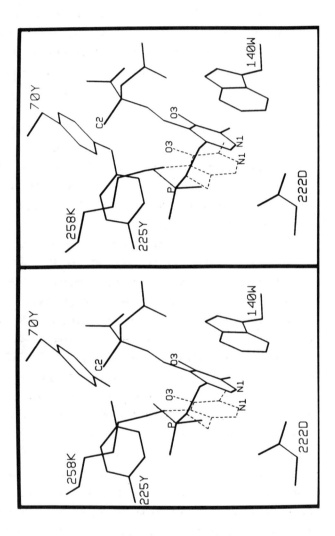

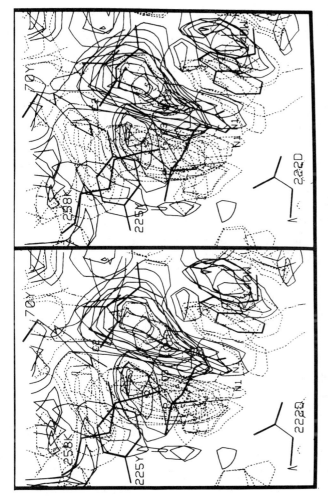

FIG. 9   TOP: Model of residues and external aldimine complex of 2-methylaspartate at the active site of subunit-1. (Tyr 70 is contributed from subunit-2). All residues are shown in their positions in the native enzyme. The internal aldimine is shown in dashed lines. BOTTOM: 3.2Å resolution methylaspartate difference density map superimposed on the atomic model. Contouring is at levels $\pm n \cdot 1.5 \cdot \sigma$ ($n = 1,2,3,...$).

nitrogen and the carboxylate side chain of Asp 222. The external aldimine carbon-nitrogen double bond is coplanar with the aromatic ring and the aldimine nitrogen is close to the phenolate oxygen. Positioning the amino acid inhibitor's α-carboxylate and side chain into density forces its α-methyl group to point perpendicular to the plane of the coenzyme/Schiff base conjugated π system. This is precisely the conformation predicted by Dunathan (41) that facilitates removal of the α-proton in the natural substrate. The enzyme's preference for the cis (maleinoid) conformation of the substrate's four carbon atoms explains the greater inhibitory effect of maleic acid over fumaric acid (43).

Formation of the external aldimine is also accompanied by repositioning of other active site residues. The side chain of Tyr 225 moves with the coenzyme ring. The indole of Trp 140, stacked with the internal aldimine ring, undergoes a similar tipping and translation. Small repositioning of the guanidinium groups of Arg 386 and 292 (from the opposing subunit) is indicated. The Lys 258 side chain, freed from its covalent bond in the internal aldimine, may be weakly interacting with the phosphate oxygen atoms. Judging from its close proximity to the external aldimine's facile proton, it seems likely that the amino group of Lys 258 may be the group catalyzing the reaction's aldimine/ketimine tautomerization (40).

Both the aldimine model and some features of the difference density image suggest that while the amino acid may stay bound in essentially the same position in the Michaelis complex, tetrahedral geminal diamine adduct, and external aldimine complexes, the coenzyme ring undergoes most of the movement. If a small amount of inhibitor were bound as a Michaelis complex (located in the same position and conformation as in the external aldimine) the low resolution difference density image at the substrate binding site would tend to be strengthened. In addition, the

presence of the Michaelis complex would decrease the level of difference density near the aromatic ring of the internal aldimine.
Both these features, due to the presence of a small amount of
Michaelis complex, are consistent with the observed difference
density (Fig. 9).

The two structures determined directly from electron density
images - those of the internal and external aldimine - as well as
the geminal diamine which was deduced from model building should
present a detailed stereochemical mechanism for the transaldimination stage of reaction.

DL-2-Hydroxymethylaspartate: Despite the additional hydroxyl
group, this compound (44) behaves like 2-methylaspartate in binding to the crystalline enzyme. The complex was obtained by soaking native enzyme crystals with the compound at 100mM. The position of the additional oxygen atom was located in a hydroxymethylaspartate (hmasp) vs methylaspartate (masp) difference map made
from |Fhmasp| - |Fmasp| structure factor amplitudes and refined iso-
morphous phases to 3.8Å resolution. The difference map shows a
single intense positive density peak at the active site at a level
10 times the RMS difference density in the map. An oxygen atom
added in proper geometry to the 2-methylaspartate:PLP external
aldimine model could be placed into the peak's center by rotation
about the single bond between the α-carbon and 2-methylene carbon
atoms (Fig. 10). This confirms both that the hydroxy compound
also forms the external aldimine complex and that the methylaspartate model was correctly fit. A small domain shift identical
to that induced by binding 2-methylaspartate was signaled by an
absence of density in the small domain region of the hmasp-masp
difference map.

L-Glutamic acid: When the natural substrate glutamate is
added at 300mM to holo-AspAT crystals, a species with a strong absorption peak at 330nm predominates. Since both the ketimine

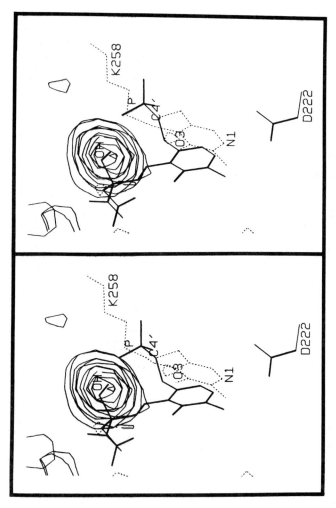

FIG. 10   Atomic model of hydroxymethylaspartate (hmasp) external aldimine complex formed by adding an oxygen atom to the methylaspartate (masp) external aldimine model.   The oxygen was centered in the large positive difference density peak in the hmasp vs masp 3.8Å resolution difference map by rotating about the C_α-C_β bond.   The map is contoured at levels of $\pm n \cdot 2.5 \cdot \sigma$ (n=1,2,3,...).

intermediate and pyridoxamine 5´-phosphate absorb in this region, absorption spectra alone cannot distinguish which species is present or cannot estimate their relative amounts.

Diffraction data from glutamate-soaked crystals and native enzyme crystals were used to construct a Fourier difference map to 3.5Å resolution. As in the 2-methylaspartate binding studies, the largest positive difference peak showed substrate binding only to subunit-1's active site and inducing a similar domain movement confined to subunit-1's small domain. Model building was used to determine whether the ketimine or a Michaelis complex with PMP was more consistent with the difference electron density image. A small rotation and translation of the coenzyme ring was needed to account for the negative and positive density features around the native enzyme's PLP model. All attempts to fit a Michaelis complex of pyridoxamine 5´-phosphate with either 2-oxoglutarate or L-glutamate failed because, when the coenzyme ring was rotated, its amino group was always positioned closer to the substrate molecule than non-bonded van der Waals contacts would allow. On the other hand, the ketimine model was easily positioned in the difference electron density. Fig. 11 shows the ketimine model with superimposed difference electron density. Thus, the difference map image along with the spectral data shows that the ketimine intermediate predominates in the crystal in the presence of 300mm glutamate.

Beta-subform: Late in the isolation and purification of the enzyme, pig heart cytoplasmic AspAT is separated into different subforms. The three most prominent forms are labeled α-, β-, and γ-subforms according to their increasing anodic mobility on starch-gel electrophoresis (45). The β-subform has about one-half the specific activity of the more abundant α-subform (which is the subject of our high resolution x-ray studies). Earlier investigators have attributed the decreased activity and charge discrepancies in the β-subform to the deamidation of an essential asparagine or glutamine residue, to the oxidation of -SH groups or the

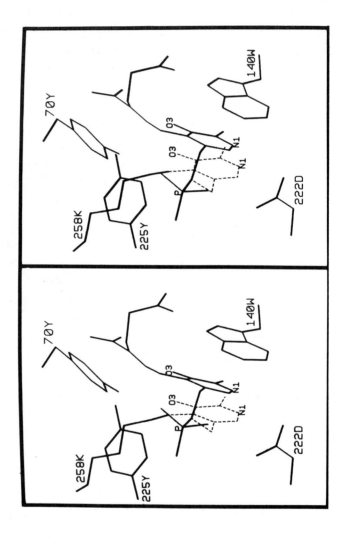

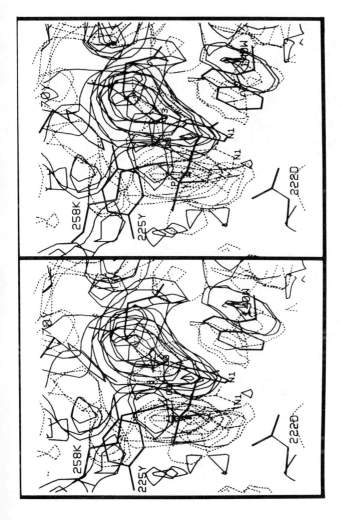

FIG. 11   TOP: Glutamate ketimine model and active site residues. The internal aldimine is shown in dashed lines. BOTTOM: 3.5Å resolution difference map formed by soaking native enzyme crystals with 300mM glutamate. Contouring is at levels of ±n·1.5·σ (n=1,2,3,...).

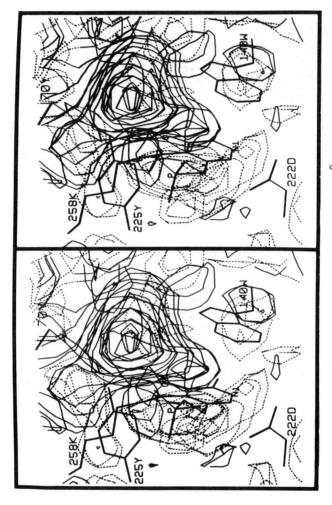

FIG. 12   β-subform difference density map at 3.8Å resolution showing the presence of an unknown substrate-like inhibitor in subunit-1's active site. Same view of the active site as in Figs. 9 and 11 with superimposed β-AspAT difference density contoured at ±n·1.5·σ (n=1,2,3,...).

coenzyme's aldehyde group, to glycosylation with acidic sugars, or to non-productive binding of the coenzyme (45,46,47,48).

Beta-AspAT crystals are isomorphous with native enzyme crystals and grow under the same conditions. A β-AspAT vs native difference map at 3.8Å resolution shows that one of the two active sites in the β-subform is chemically modified by a substrate-like inhibitor. The enzyme dimer crystallizes so that the inhibitor occupies only the active site of subunit-1 and that only sub- unit-1's small domain is shifted. Stereo Fig. 12 shows the active site region with superimposed difference electron density. A tilting of the coenzyme ring and movements of several active site residues are indicated. At the resolution in this map, it is impossible to characterize the inhibitor other than to estimate its approximate size and shape. When the ketimine model is super- imposed in the β-subform difference density, the inhibitor is shown to be roughly the same size or slightly larger than the natural substrates. Because both Arg 386 and Arg 292 side chains are interacting with the β-inhibitor in much the same way that they bind other compounds, we suspect that the inhibitor may also be a dicarboxylic acid.

Now that the x-ray studies have shown that β-AspAT is blocked at one of its active sites, attempts to isolate and chemically identify the inhibitor are now underway (49). This work may indicate how AspAT is regulated in vivo by undergoing a suicide reaction with a naturally occurring compound.

COMPLEMENTARY USE OF CRYSTALLOGRAPHY AND SPECTROSCOPY

The coenzyme chromophore in aspartate aminotransferase has been long recognized as a sensitive "reporter" of the coenzyme's covalent state during the various reaction stages. Recent studies of the formation of substrate and inhibitor complexes in the crystalline enzyme have relied on the chromophore both to identify intermediate complexes and to characterize resulting structural changes. Direct experimental evidence that the coenzyme is re-

oriented came from polarized light spectra of cAspAT crystals
(50). This final section will outline how ultraviolet/visible
polarized light spectroscopy is being used in a complementary way
with x-ray crystallography to study the binding of substrates and
inhibitors in AspAT crystals.

AspAT Crystal Ultraviolet/Visible Spectra

The vitamin-B_6 coenzyme and its reaction intermediate complexes in AspAT have well-characterized ultraviolet and visible
spectral properties (summarized in 1,2). The protonated internal
aldimine nitrogen at low pH has a 430nm absorption band responsible for the enzyme's deep golden-yellow color. Deprotonation of
the Schiff base with a pKa of 6.3 gives rise to a 363nm absorption
band. The external aldimine has similar properties, except that
its pKa has been substantially shifted well above 6.3 so that it
remains protonated throughout the pH range at which the enzyme is
stable (1). An intense absorption at 490nm, formed when the
crystals are soaked with the compound L-erythro-β-hydroxyaspar-
tate, turns the crystals light pink and indicates the presence of
the stablized carbanion or quinonoid species. All of the amino
coenzyme forms (ketimine, pyridoxamine 5´-phosphate, and geminal
diamine adduct) are characterized by absorptions centered at
330nm.

While spectral properties of all these complexes are believed
to be identical in solution and in the crystal, the absorption
spectra recorded in pig heart cAspAT crystals can sometimes vary
from those in solution because of the crystal's half-sites reactivity. The presence of unreactive internal aldimine in sub-
unit-2's active site can complicate the resulting crystal spectra.

Linear Dichroism of AspAT Crystals

Polarized light passed through an oriented sample often shows
varying absorptions when the beam is aligned along different

directions. This phenomenon called "linear dichroism" (reviewed in 51) is greatest in crystalline specimens where chromophores are spatially positioned by the crystal lattice. A linear dichroic measurement is usually expressed as the ratio of the extinctions recorded with the incident light beam's electric vector aligned along two defined directions. Optical axes coinciding with crystallographic axes serve as reference directions in crystalline samples. Since the magnitude of the dichroic behavior is very sensitive to the chromophore's orientation, any movement of the chromophore can be detected by changes in the polarization ratio. Discussed in this section are the methods being used to correlate AspAT crystal polarized light spectra with the reorientations of AspAT's coenzyme chromophore that have been observed crystallographically.

The experimental techniques used to record polarized light spectra of AspAT single crystals mounted in the beam of a conventional UV/Vis spectrophotometer or on the stage of a high-quality microspectrophotometer have been detailed in earlier papers (52,53). In the cytoplasmic AspAT crystals where spectra are recorded with the incident light beam perpendicular to the prominent "B" or [0,1,0] face of the plate-like crystal, the polarization ratio is expressed as the extinction recorded along the c-crystal axis divided by the extinction along the a-axis. A precise alignment of the polarizer with these two crystal axes is simplified by the orthorhombic symmetry since maxima or minima in the extinctions occur along these axes. Recording spectra along the b-axis is difficult because of the crystal thickness in the a- and c-dimensions.

Fig. 13 shows a native enzyme crystal's polarized light absorption spectrum recorded along both the c- and a-crystal axes. The overlapping component absorption bands giving rise to the complex spectra are resolved in order to get more accurate estimations of the band heights and polarization ratios. The spectrum

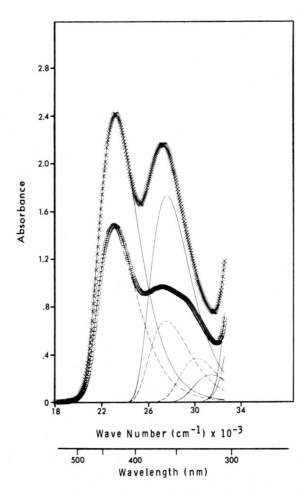

FIG. 13   Polarized light ultraviolet/visible absorption spectra
of orthorhombic cAspAT crystals.   The spectra are recorded with
the incident light beam's electric vector aligned along the crystallographic c-axis (above, X) and a-axis (below, O).   The complicated spectra have been resolved into component absorption
bands by least-squares fitting to a series of overlapping log-
normal distributions (solid curves for C-polarization, dashed
curves for A-polarization) (Spectral data courtesy of Marvin
Makinen).

shown in Fig. 13 has been resolved by fitting each individual band to a single log-normal distribution (54).

The visible and near-ultraviolet absorptions in the AspAT spectra arise from low-lying π-π^{*} transitions in the aromatic ring of the coenzyme chromophore. As in most aromatic systems, the transition dipole moments giving rise to these different bands are thought to be strongly polarized in the plane of the π-system. The transition dipole moment (TDM) is a quantum-mechanical vector integral whose orientation corresponds to a direction of maximum absorption and whose square determines the strength of the absorption (51). Before a quantitative correlation can be made between changes in chromophore orientation and changes in the polarized light spectra, it is necessary to first determine each absorption band's transition dipole moment orientation. For many of the absorption bands in the AspAT spectrum it may be possible to find the TDM orientations by comparing measured polarization ratios with those predicted from the crystallographically-determined coenzyme ring planes.

The smooth curves in Fig. 14 show how the C/A polarization ratio would theoretically vary with different orientations of the transition dipole moment within the coenzyme ring. In these calculations the TDM is treated simply as a double-headed vector lying within the plane of the coenzyme's pyridine ring and orientated relative to atoms in the ring by the angle θ (defined in Fig. 15). The theoretical curves are generated by varying the orientation of the TDM vector within each aromatic chromophore's ring plane and calculating the expected polarization ratios. In general, these curves yield two degenerate TDM orientations giving the same calculated polarization ratio. Other information is often needed to decide which of the two orientations is correct.

The equations used to calculate theoretical C/A polarization ratios for each absorption band (derived in 51) are simple for the

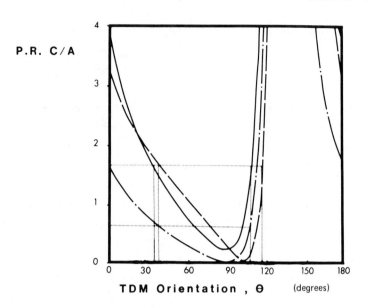

FIG. 14   Theoretical dependence of the C/A polarization ratio as a function of all possible orientations of the 430nm absorption transition dipole moment (TDM). Curves for the native enzyme, β-subform, and methylaspartate complex are calculated from the crystallographically-determined chromophore ring planes (see text). In general, these curves provide two possible TDM orientations from the spectroscopically-measured polarization ratios (dotted lines).

orthorhombic system:

$$(C/A) \ P.R. \ = \ \frac{\sum\limits_{i=1}^{n} 3\bar{\varepsilon} \cdot \cos^2(\mu_i \cdot c)}{\sum\limits_{i=1}^{n} 3\bar{\varepsilon} \cdot \cos^2(\mu_i \cdot a)}$$

where $\mu_i \cdot c$ and $\mu_i \cdot a$ are the angles between the transition dipole moment vectors in each of the n unique chromophore orientations

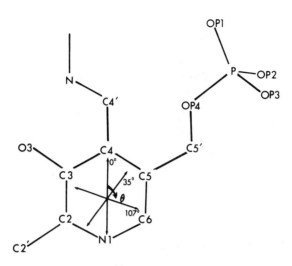

FIG. 15   Definition of the angle θ giving the orientation of the transition dipole moment in the planar ring of the coenzyme chromophore. The angle θ is measured as a positive clockwise rotation when viewed from the ring's A face. TDM orientations at 35° and 107° are shown.

and the $\underline{c}$- and $\underline{a}$-axes, respectively, and $\bar{\varepsilon}$ is the mean solution extinction coefficient. When only one type of chromophore is present, the extinction coefficients cancel in the equation.

    The discussion below, dealing only with the 430nm band, illustrates how the orientation of the various transition dipole moments may be pinpointed. The internal aldimine 430nm band transition dipole orientation, estimated from both the native enzyme and β-subform crystal linear dichroic measurements, agrees well with quantum-mechanical calculations.

    Native Enzyme Internal Aldimine Complex: In the native PLP-holoenzyme crystal at low pH, the protonated internal aldimine gives rise to the 430nm absorption band. Although there are eight chromophores in the unitcell (one in each of the two subunits of four asymmetric units), only the two distinct classes of any one asymmetric unit are used for the summation. The equation for the predicted C/A polarization ratio as a function of theta for the native enzyme 430nm bands is:

$$\text{(C/A) P.R. } (\theta) = \frac{\cos^2(\mu_{\theta,1} \cdot c) + \cos^2(\mu_{\theta,2} \cdot c)}{\cos^2(\mu_{\theta,1} \cdot a) + \cos^2(\mu_{\theta,2} \cdot a)}$$

where $\mu_{\theta,n} \cdot a$ or $\mu_{\theta,n} \cdot c$ denote the angle between the TDM vector, μ_θ and the a- or c-axes when the TDM is oriented within the plane of the coenzyme ring at an angle θ. The subscript denotes the sub-unit number. The angles $\mu_\theta \cdot a$ and $\mu_\theta \cdot c$ are measured geometrically using the coenzyme ring planes determined by model fitting to electron density.

The above equation, plotted as curve #1 of Fig. 14, shows the calculated polarization ratio for all possible orientations of the TDM within the coenzyme rings. A more qualitative interpretation can be seen from Fig. 16 (top) where the pyridoxal rings from both active sites are shown projected onto the crystal's B-face in the same orientation seen in the linear dichroism experiments. Notice that if the transition dipole moments were oriented at $\theta = 90°$ (with both nearly parallel to the a-crystal axis) the absorption along the c-axis would be at a minimum and the calculated C/A polarization ratio would approach zero. Likewise, if the transition dipoles were oriented at about 130° (both nearly parallel to the a-axis), the absorption along the a-axis would approach zero and the calculated polarization ratio would become very large. These predicted minima and maxima in the polarization ratio can be seen in the calculated curve #1 of Fig. 14.

Using the native crystal's 430nm band observed polarization ratio of 1.60, curve #1 suggests that the TDM may lie at either $\theta = 35°$ or 107°. These two possible TDM orientations are shown projected in the coenzyme ring planes in Figs. 15 and 16.

AspAT β-subform: Since the unknown coenzyme/inhibitor complex occupying the active site of subunit-1 in crystalline β-AspAT lacks any 430nm absorption, it is assumed that the absorption is due only to the protonated internal aldimine of subunit-2. Judging from the β-AspAT vs native difference map, the orientation of the PLP ring in subunit-2 (Fig. 16, center) is the same as in the

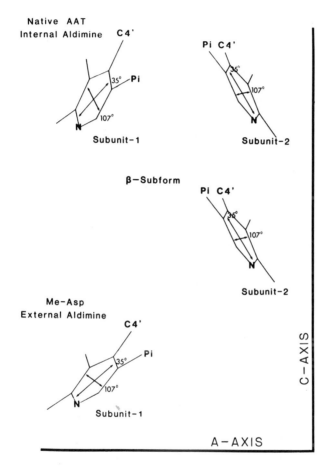

FIG. 16   Projections of the coenzyme rings onto the crystal's B or [0,1,0] face used in the determination of the 430nm band TDM orientation.   The coenzyme rings having 430nm absorptions in the native enzyme, β-subform, and high pH methylaspartate external aldimine are shown at top, center, and bottom, respectively.   TDM vectors oriented at 35° and 107° are indicated.

native enzyme. The equation used to calculate curve #2 of Fig. 14
requires only the position of the coenzyme in subunit-2:

$$(C/A) \text{ P.R. } (\theta) = \frac{\cos^2(\mu_{\theta,2} \cdot c)}{\cos^2(\mu_{\theta,2} \cdot a)}$$

Using the polarization ratio of 1.62 measured in β-AspAT
crystals and curve #2 of Fig. 14, the 430nm transition dipole is
estimated to lie at either θ = 38° or 116°.

Quantum-Chemical Calculations: Calculations by F. Savin
(55,56) using the all-valence-electrons CNDO/S method indicated
that the 430nm transition dipole moment is oriented approximately
at θ = 40°. This result breaks the ambiguity between the two
degenerate orientations and shows that the true orientation of the
430nm transition dipole probably lies between the experimentally
estimated values of 35° and 38°, rather than between 107° and
116°.

2-Methylaspartate Complex: Crystals soaked with 2-methyl-
aspartate are bright yellow and have strong 430nm absorptions over
a wide pH range. So far, we have not been able to quantitatively
correlate the available polarized light spectra of these crystals
at pH 5.4 with the crystallographic results. As discussed earlier, soaking AspAT crystals at pH 5.4 with this inhibitor causes
the external aldimine complex in subunit-1's active site to form,
while subunit-2 remains unreactive as the internal aldimine. A
small amount of Michaelis complex may also be present in sub-
unit-1. Because these three complexes all have 430nm absorptions
bands and because their relative concentrations are unknown, it
becomes difficult to predict the spectral results from the crystallographic image.

The most straightforward analysis can be done with the 300mM
2-methylaspartate complex at a higher pH. Above pH 7 or 8, where
the free internal aldimine of subunit-2 and Michaelis complex of
subunit-1 are titrated to their 363nm-absorbing form, the only

species with a 430nm absorption in the crystal is the external aldimine complex of subunit-1. Because the pKa for the external aldimine is substantially higher than that of the internal aldimine, the external aldimine remains fully protonated at high pH.

Although the external aldimine complex has not yet been observed at high pH crystallographically, we expect its orientation to be nearly identical to that observed in the low pH Fourier difference maps (shown projected onto the B crystal face in Fig. 16, bottom). This ring orientation was used to calculate curve #3 of Fig. 14. Since the external and internal aldimine complexes are isoelectronic and structurally similar, their TDM magnitudes and orientations may be assumed to be identical. Curve #3 predicts the 430nm absorption band's C/A polarization ratio to be 0.65, 0.60, and 0.56 using the earlier estimates of the TDM vectors at $\theta=35°$, 38°, and 40°, respectively. This agrees very well with polarization ratios of about 0.6 that have previously been observed above pH 7 in several methylaspartate-soaked crystals. However, a recent (presumably more accurate) microspectrophotometer measurement suggested that the polarization may go as low as 0.4. To account for a polarization ratio of 0.4 at high pH, the tilt of the external aldimine ring would have to increase an additional 12° about the C5´-C5 bond.

The accuracy of these calculations depends to large extent on how well the coenzyme rings' spatial orientations can be determined from x-ray crystallography. The current ring positions are probably fit to within ±5° of their true positions. Although the effect of a small positional error on the calculated polarization ratios is complicated and difficult to represent, a random 5° error in the tilt of the plane of the ring causes an approximate 5° change in the angle θ. Considering all possible sources of error in the crystallographic interpretation and in the polarization ratios, it should be possible to locate the transition dipole moments to within ± 5° of their true positions.

Linear Dichroism as an Approach to the Problem of Mixtures of
Intermediates

Before the binding of a substrate or inhibitor can be completely characterized, the level of minor species must be quanti-
tated. This is especially important in crystallographic refinement where the occupancies of different complexes must be known or
estimated. While the substrates and inhibitors bound in crystalline AspAT usually seem to form one of several possible complexes,
other complexes may be present in minor amounts. Even 2-methyl-
aspartate (expected to form only the external aldimine complex)
forms a small amount of Michaelis complex. Thus, a fundamental
problem in our study of AspAT substrate complexes is to measure or
estimate relative concentrations in mixtures of intermediates.

X-ray crystallography cannot always be used to quantify
mixtures since different species bound simultaneously at the same
site creates a confusing electron density image. This is especially true in low resolution difference electron density maps.
Likewise, spectroscopic methods cannot always be used by themselves to quantify reaction intermediates since several different
complexes may have identical spectral properties. However, in
cAspAT crystals, it may be possible to use polarized light spectra
together with crystallography to quantify these species. Because
much of the needed spectral and crystallographic data has not yet
been collected, the procedure described below is only an outline
of an approach that might be used. The discussion will focus on
quantifying the relative amounts of ketimine complex and pyridoxamine 5´-phosphate formed in subunit-1's active site by soaking
native AspAT crystals with 300mM L-glutamate.

Because the enzyme crystals are catalytically competent,
adding substrate will yield an equilibrium mixture comprised of
all the substrate, product, and intermediary complexes shown in
Fig. 7. In the conditions used in these experiments, however,
comparatively few complexes predominate to any measurable extent.

As mentioned before, glutamate reacts in the crystal primarily in one of the two active sites to form species with strong absorptions at 330nm. The amount of substrate bound in the active site of subunit-2 is negligible. A very low level of difference density at the coenzyme ring in subunit-2 suggests that a small portion (probably less than 10%) of the interal aldimine complex is slowly transaminated to pyridoxamine 5´-phosphate. A 430nm absorption still present in the glutamate-soaked crystals comes, therefore, from the unreacted internal aldimine of subunit-2, although any external aldimine in subunit-1 would also contribute if present. Because the observed C/A polarization ratio is 1.58 in the 430nm band, we believe that virtually no external aldimine complex is present. The P.R. value of 1.58 is essentially the same as value of 1.62 seen in β-AspAT crystals where only the internal aldimine of subunit-2 has a 430nm absorption. If increasing amounts of external aldimine complex were present in subunit-1, the predicted polarization ratio would decrease to about 0.9.

If all the 430nm absorption comes from the unreactive subunit-2, the species occupying the active site of subunit-1 must be responsible for the large 330nm absorption band. Therefore, the species present must be either a Michaelis complex of the substrate with PMP, the ketimine complex, or a mixture of the two. Unfortunately, since both species are characterized by 330nm absorptions (1,50), the two possibilities cannot be distinguished by their absorption spectra alone. Model building showed the ketimine complex to be more consistent with the difference density image. The negative and positive density features on different sides of the internal aldimine's ring position indicated that the coenzyme ring in the ketimine complex tilts into a position much like that of external aldimine. Because the ketimine model fit very well, and since bad steric conflicts were found when a Michaelis complex of a "tipped" PMP and substrate molecule were fit to the density, we concluded that the major species present is the ketimine complex. A small amount of Michaelis complex would be consistent with the difference density, however, if the substrate has the same position as in the ketimine intermediate and if the coenzyme ring of PMP is not tipped. The problem is to estimate the relative amounts of ketimine and glutamate/PMP Michaelis complex formed in the active site of subunit-1 in the 300mM glutamate soak. Assuming that the data are sufficiently accurate and that the polarization ratio changes are sufficiently large, it should be possible to find their relative amounts using polarized light spectra as follows. One must first make the assumption that the 330nm transition dipole moments are of the same orientation and magnitude in both the ketimine complex and pyridoxamine 5´-phosphate. Because these two species are iso-electronic and chemically similar, this should be a valid assumption. The 330nm TDM orientation may be found either from the crystalline PMP enzyme form or from the borohydride-reduced native enzyme by comparing the measured C/A polarization ratios with the theoretical polarization ratios derived from the crystallographically-determined ring orientations.

The observed polarization ratio of the 330nm band in crystals soaked with 300mM glutamate should be given by the following equation:

$$(C/A) \ P.R. = \frac{r \cdot \cos^2(\mu_M \cdot c) + (1-r) \cdot \cos^2(\mu_K \cdot c)}{r \cdot \cos^2(\mu_M \cdot a) + (1-r) \cdot \cos^2(\mu_K \cdot a)}$$

where the subscript M or K refers to the Michaelis complex or ketimine, respectively.

The angles between the TDM and crystal axes ($\mu \cdot a$ or $\mu \cdot c$) can be found using the ketimine and Michaelis complex coenzyme ring orientations and the assumed orientation of the TDM within the coenzyme ring. The equation can be solved for the value r which is the ratio of the concentration of Michaelis complex to ketimine.

Thus, although the PMP Michaelis complex and ketimine cannot be distinguished by their spectral properties alone, their differences in orientation can be used to quantify their relative amounts. Whether this kind of approach will be useful in quantifying these and other mixtures of intermediates will have to await the further collection of accurate spectral and crystallographic data.

ACKNOWLEDGEMENTS

We thank Carol Galbraith for preparing this manuscript and Amy Lilienfeld for technical proofreading. This work has been supported by U.S. Public Health Service (NIH) grants AM-01549 and AM-17563.

REFERENCES

1. Braunstein, A.E., in The Enzymes, Boyer, P.D., ed., 3rd Ed., Academic Press, New York, (1973) 379.

2. The Transaminases, Christen, P. & Metzler, D.E., eds, John Wiley & Sons Inc., New York, in press.

3. Borisov, V.V., Borisova, S.N., Kachalova, G.S., Sosfenov, N.I., Voronova, A.A., Vainshtein, B.K., Torchinsky, Yu.M., Volkova, G.A., & Braunstein, A.E., Dokl. Akad. Nauk SSSR 235, 212 (1977).

4. Borisov, V.V, Borisova, S.N., Kachalova, G.S., Sosfenov, N.I., Vainshtein, B.K., Torchinsky, Yu.M., & Braunstein, A.E., J. Mol. Biol. 125 275 (1978).

5. Borisov, V.V., Borisova, S.N., Sosfenov, N.I., & Vainshtein, B.K., Nature 284, 189 (1980).

6. Harutyunyan, E.G., Malashkevich, V.N., Tersyan, S.S., Kochkina, V.M., Torchinsky, Yu.M., & Braunstein, A.E. FEBS Letters 138, 113 (1982).

7. Borisov, V.V., Borisova, S.N., Kachalova, G.S., Sosfenov, N.I., & Vainshtein, B.V., in The Transaminases, Christen, P. & Metzler, D.E., eds., John Wiley & Sons Ins., New York, in press.

8. Harutyunyan, E.G., Malashkevich, V.N., Kochkina, V.M., & Torchinsky, Yu.M., in The Transaminases, Christen, P. & Metzler, D.E., eds., John Wiley & Sons Inc., New York, in press.

9. Eichele, G., Ford, G.C., Glor, M., Jansonius, J.N., Mavrides, C., & Christen, P., J. Mol. Biol. 133, 161 (1979).

10. Ford, G.C., Eichele, G., & Jansonius, J.W., Proc. Natl. Acad. Sci. USA 77, 2559 (1980).

11. Jansonius, J.N., Eichele, G., Ford, G.C., Picot, D., Thaller, C., & Vincent, M.G., in The Transaminases, Christen, P. & Metzler, D.E., eds., John Wiley & Sons Inc., New York, in press.

12. Arnone, A., Rogers, P.H., Schmidt, J., Han, C.-N., Harris, C.M., & Metzler, D.E., J. Mol. Biol. 112, 509 (1977).

13. Arnone, A., Briley, P.D., Rogers, P.H., Hyde, C.C., Metzler, C.M., & Metzler, D.E., in Molecular Structure and Biological Activity, Griffin, J.F. & Duax, W.L., eds., Elsevier North-Holland Inc., New York (1982) 57.

14. Arnone, A., Rogers, P.H., Hyde, C.C., Briley, P.D., Metzler, C.M., & Metzler, D.E. in The Transaminases, Christen, P. & Metzler, D.E., eds, John Wiley & Sons Inc., New York, in press.

15. Braunstein, A.E., & Shemyakin, M.M., Doklady AN SSSR 85, 1115 (1953a).

16. Braunstein, A.E., & Shemyakin, M.M., Biokhimiya 18, 393 (1953b).

17. Metzler, D.E., Ikawa, M., & Snell, E.E., J. Am. Chem. Soc. 76, 648 (1954).

18. Ovchinnikov, Yu.A., Egorov, C.A., Aldanova, N.A., Feigina, M.Yu., Lipkin, V.M., Abdulaev, N.G., Grishin, E.V., Kiselev, A.P., Modyanov, N.N., Braunstein, A.E., Polyanovsky, O.L., & Nosikov, V.V., FEBS Letters 29, 31 (1973).

19. Doonan, S., Doonan, H.J., Hanford, R., Vernon, C.A., Walker, J.M., da S. Airoldi, L.P., Bossa, F., Barra, D., Carloni, M., Fasella, P., & Riva, F., Biochem. J. 149, 497 (1975).

20. Blundell, T.L. & Johnson, L.N., Protein Crystallography, 1st edn., Academic Press, New York/London/San Francisco (1976).

21. Blow, D.M. & Crick, F.H.C., Acta Cryst. 12, 794 (1959).

22. Eisenberg, D., in The Enzymes, Boyer, P.D., eds., Vol. 1, 3rd ed., Academic Press, New York/London/San Francisco (1970) 1.

23. Bricogne, G., Acta Cryst. A32, 832 (1976).

24. Hendrickson, W.A. & Ward, K.B., J. Biol. Chem. 252, 3012 (1977).

25. Wilson, I.A., Skehel, J.J., & Wiley, D.C., Nature 289, 366 (1981).

26. Murthy, M.R.N., Reid, T.J., Sicignano, A., Tanaka, N., & Rossmann, M.G., J. Mol. Biol. 152, 465 (1981).

27. Honzatko, R.B., Crawford, J.L., Monaco, H.L., Ladner, J.E., Edwards, B.F.P., Evans, D.R., Warren, S.G., Wiley, D.C., Ladner, R.C., & Lipscomb, W.N., J. Mol. Biol. 160, 219 (1982).

28. Zapatero, C.A., Meguid, S.S.A., Johnson, J.E., Leslie, A.G.W., Rayment, I., Rossman, M.G., Suck, D., & Tsukihara, Nature 286, 33 (1980).

29. Liljas, L., Unge, T., Jones, T.A., Fridborg, K., Lovgren, S., Skoglund, U., & Strandberg, B., J. Mol. Biol. 159, 93 (1982).

30. Hendrickson, W.A. & Lattman, E.E., Acta Cryst. B26, 136 (1970).

31. Hendrickson, W.A., Acta Cryst. B27, 1572 (1971).

32. Ramachandran, G.N., Ramakrishnan, C., & Sasisekharan, V., J. Mol. Biol. 7, 95 (1963).

33. Alber, T., Petsko, G.A., & Tsernoglou, D., Nature 263, 297 (1976).

34. Plapp, B.V., Eklund, H., & Branden, C.I., J. Mol. Biol. 122, 23 (1978).

35. Eklund, H., Plapp, B.V., Samama, J.P., & Branden, C.I., J. Biol. Chem. 257, 14349 (1982).

36. Rees, D.C., Lewis, M., Honzatko, R.B., Lipscomb, W.N., & Hardman, K.D., Proc. Natl. Acad. Sci. USA 78, 3408 (1981).

37. Kirsch, J.F. & Julin, D.A., Fed. Proc. 41, 628 (1982).

38. Boettcher, B., & Martinez-Carrion, M., Biochemistry 15, 5657 (1976).

39. Schlegel, H., Zaoralek, P.E., & Christen, P., J. Biol. Chem.
    252, 5835 (1977).

40. Ivanov, V.I. & Karpeisky, M. Ya., Adv. Enzymol. 32, 21 (1969).

41. Dunathan, H.C., Adv. Enzymol. 35, 79 (1971).

42. Melander, W.R., FEBS Letters 51, 5 (1975).

43. Velick, S.F. & Vavra, J., J. Biol. Chem. 237, 2109 (1962).

44. Walsh, J.J., Metzler, D.E., Powell, D., & Jacobson, R.A., J.
    Am. Chem. Soc. 102, 7136 (1980).

45. Martinez-Carrion, M., Turano, C., Chiancone, E., Bossa, F.,
    Giartosio, A., Riva, F., & Fasella, P., J. Biol. Chem. 242,
    2397 (1967).

46. Denisova, G.F. & Polyanovsky, O.L., FEBS Letters 35, 129
    (1973).

47. John, R. & Jones, R., Biochem. J. 141, 401 (1974).

48. Williams, J.A. & John, R.A., Biochem. J. 177, 121 (1979).

49. Metzler, C.M. in Chemical and Biological Aspects of Vitamin
    B₆ Catalysis, Evangelopoulos, ed., Alan R. Liss, New York,
    1983, in press.

50. Metzler, C.M., Metzler, D.E., Martin, D.S., Newman, R.,
    Arnone, A., & Rogers, P., J. Biol. Chem. 253, 5251 (1978).

51. Hofrichter, J. & Eaton, W.A., Annu. Rev. Biophys. & Bioeng.
    5, 511 (1976).

52. Metzler, C.M., Rogers, P.H., Arnone, A., Martin, D.S., &
    Metzler, D.E., in Methods in Enzymology 62, 551 (1979).

53. Metzler, D.E., Metzler, C.M., Likos, J.J., Ueno, H.,
    Feldhaus, R., Scott, R.D., Martin, D.S., Arnone, A., Briley,
    P.D., Rogers, P.H., Hyde, C.C., & Makinen, M., in
    Physico-chemical Bases of Enzymatic Catalysis and its
    Regulation, Torchinsky, Yu.M., ed., Nauka, Moscow, USSR, in
    press.

54. Siano, D.B. & Metzler, D.E., J. Chem. Physics 51, 1956
    (1969).

55. Braunstein, A.E. in Frontiers of Bioorganic Chemistry and Molecular Biology, Ananchenko, S.N., ed., Pergamon Press, New York (1980).

56. Savin, F. in Physico-chemical Bases of Enzymatic Catalysis and its Regulation, Torchinsky, Yu.M., ed., Nauka, Moscow, USSR, in press.

HEME ENZYME STRUCTURE AND FUNCTION

Thomas L. Poulos* and Barry C. Finzel*

Department of Chemistry

University of California, San Diego

La Jolla, California

ABSTRACT

The current status of heme enzyme structure and function is examined with emphasis placed on results from protein crystallography. Wherever possible, these results are correlated with information obtained from other experimental and theoretical approaches. Comparative structural analyses are made between various heme proteins in an effort to identify structural differences responsible for functional diversity. Lastly, based upon both x-ray structures and other available information, we review facts and speculations regarding heme enzyme catalyzed fission of the O-O bond and possible mechanisms of interprotein electron transfer reactions.

I. GENERAL INTRODUCTION

Heme proteins are essential components in a variety of physiological processes ranging from the simplest prokaryotes to man. A few examples are photosynthetic, respiratory, and microsomal electron transfer, and the utilization and metabolism of oxygen and peroxides. While a large number of heme proteins participate in these processes, they naturally fall into three functionally distinct groups: electron carriers

---

*Current affiliation: Protein Engineering Division, Genex Corporation, Gaithersburg, Maryland.

(cytochromes), reversible O_2 binders (globins) and heme enzymes. Each group is functionally unique yet they all share at least one common and important structural feature: an iron porphyrin bound at the active center. A long sought after goal of heme protein biochemistry is to understand how interactions between the protein and iron porphyrin determine function. By doing so, we can begin to understand how Nature extracts functional diversity from an important common structural feature.

Much has been written about the structure and function of globins and cytochromes in recent years due in large part to the wealth of crystallographic data available. However, little has been said about heme enzymes because, until very recently, little was known about the detailed three dimensional structure of any heme containing catalyst. Structural information for cytochrome *c* peroxidase (CCP) and catalase, and preliminary data on cytochrome P450 are now available which indicate both structural similarities and differences relevant to the function of these enzymes.

In this article we will consider detailed hypothetical models of two processes common to many heme enzymes: 1) the mechanism of O-O bond cleavage and 2) mechanisms of interprotein electron transfer reactions. The discussion is based upon structural information obtained through x-ray crystallography coupled with results from other experimental and theoretical treatments. Our emphasis will be on cytochrome *c* peroxidase, primarily because the structure has been refined at high resolution but also because the peroxidase is an excellent model system for understanding both O-O bond cleavage and intermolecular electron transfer reactions.

II. INTRODUCTION TO HEME ENZYMES

Heme enzymes can be categorized into three types; peroxidases, oxygenases, and oxidases. Peroxidases catalyze heterolysis of the RO-OH bond (R = H or an alkyl group) and store both oxidizing equivalents of the peroxide molecule within the enzyme active center for utilization in the oxidation of a variety of substrates ranging from halogens to other proteins. Oxygenases cleave the O-O bond of molecular oxygen and then insert either one or both O_2 derived oxygen atoms into an aliphatic or aromatic molecule. Those that insert one oxygen atom are called monoxygenases and those that insert two, dioxygenases. Oxidases carry out the 4-electron reduction of O_2 to H_2O and serve as the terminal oxidase in respiratory electron transport.

a. Peroxidases

The most thoroughly studied peroxidases are those from horseradishes, turnips, and yeast mitochrondria. These peroxidases share common molecular properties; each is a single polypeptide chain of 30,000-40,000 daltons, each contains a single, noncovalently bound ferric protoporphyrin IX, and the active site sequences exhibit strong homologies(1-5).

A generalized catalytic cycle shared by all peroxidases is outlined in Scheme I.

The substrate is shown as a hydroperoxide, ROOH, since peroxidases are capable of utilizing peroxides where the R group is an alkyl moiety, though H_2O_2 is preferred. X represents an active site group which serves as the source of one reducing equivalent in the reaction with peroxides.

The first step in the reaction cycle involves coordination of RO-OH with the heme iron atom giving a Michaelis-Menten type complex. A true ES complex has never been directly observed since the reaction between the enzyme and peroxides is so fast. Existence of a prereaction complex is inferred, nevertheless, from the kinetic data(6). Second, the enzyme catalyzes heterolysis of the RO-OH bond releasing ROH while retaining one peroxide derived oxygen atom. As shown in Scheme I, the enzyme bound oxygen atom has been reduced to the level of H_2O by removing one electron from the iron atom, giving Fe(IV), and one electron from an active site X group generating an X• radical. We will consider the nature of X• further in Section IV.

After formation of this Fe(IV)X• intermediate (called Compound I) in the second step of the reaction, the third and fourth steps involve a one electron reduction of Compound I to Compound II followed by another one electron reduction of Compound II to regenerate the native enzyme.

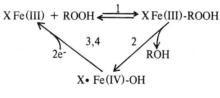

Scheme I

The type of molecule serving as the reductant of Compound I determines the precise physiological role of any particular peroxidase. Horseradish peroxidase oxidizes aromatic molecules including indole plant growth hormones[7]. Halogens are also oxidized[8], with the best known halogen oxidizing peroxidase being chloroperoxidase from the slime mold *Caldariomyces fumago*[9]. Another interesting halogen oxidizing peroxidase is myeloperoxidase found in mammalian leukocytes[10]. During infection, stimulation of leukocytes results in an oxidative burst and the production of H_2O_2[11, 12]. H_2O_2 and Cl^- ions then react with myeloperoxidase to generate a potent antibacterial agent, hypochlorous acid (HOCl), the active ingredient in common household bleach[13-17]. The list of peroxidases continues, including those involved in thyroxine metabolism[18, 19], the formation of melanin[20], and prostoglandin metabolism[21]. In our own laboratory, we have studied the structure of cytochrome c peroxidase (CCP) which oxidizes cytochrome c in yeast mitochondria.

Catalase is a highly specialized peroxidase in that H_2O_2 is both the oxidant of the native enzyme and the reductant of Compound I. Therefore, the overall catalase reaction is $H_2O_2 \rightarrow O_2 + H_2O$. As we shall see, this specialization is reflected in considerable differences between the structure of catalase and other peroxidases.

b. Oxygenases

Like peroxidases, oxygenases are ubiquitous and function in a number of different physiological processes. We focus our attention on the monoxygenase which has received the most attention, cytochrome P450. The intense interest in P450 stems primarily from its participation in the detoxification of xenobiotic aromatic hydrocarbons[22], the conversion of these substances to carcinogenic epoxides[22], and steroid metabolism[23, 24]. Also of interest is the ability of xenobiotics to induce the production of various microsomal P450s, thereby offering an excellent system for studying the genetics and regulation of an important detoxification system[25].

Even though there exist a large number of constituitive and inducible P450s, there are basically only two types of P450 monoxygenase systems, microsomal and bacterial/mitochondrial[26]. Differences in the two systems are outlined in Scheme II. In microsomes, electrons are sequentially transferred from NADH to an FAD/FMN flavoprotein and then to P450. In mitochondria and bacteria a third component, a ferredoxin-like two iron-two sulphur protein, mediates the transfer of

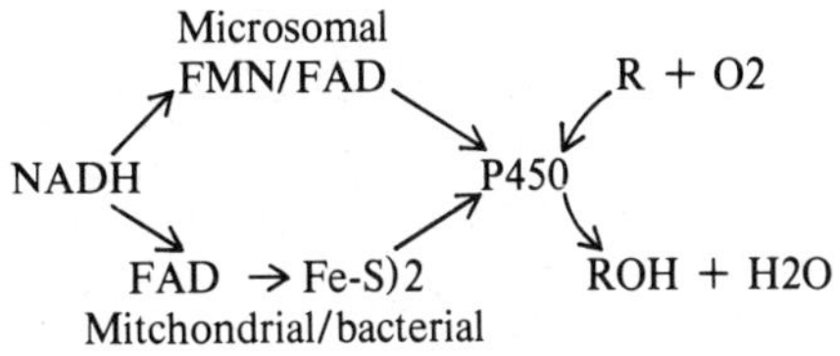

Microsomal
FMN/FAD   R + O2

NADH   P450

FAD → Fe-S)2   ROH + H2O
Mitchondrial/bacterial

Scheme II

electrons from the flavoprotein. In this case the flavoprotein contains only FAD. The types of substrates hydroxylated in various systems also differ. Liver microsomal P450 hydroxylates a variety of xenobiotic hydrocarbons as the first step in the detoxification and elimination of these substances(27-29). Adrenal cortex mitochondrial P450s are involved in steroid metabolism(30-35), while bacterial P450s initiate the breakdown of organic molecules used as the primary carbon source for growth(36).

With the exception of all components of bacterial systems, the eukaryotic flavoprotein and P450 are associated with membranes. Nevertheless, all P450s share common molecular properties. For example, all are single polypeptide chains of 45,000-55,000 daltons, contain a single ferric protoporphyrin IX, exhibit similar spectral properties upon binding ligands and substrates, and unlike most other heme proteins, one of the axial ligands is cysteine(37-50). The complete sequence of one mammalian(51) and one bacterial P450(52) are known, along with partial data for others(53-55). There are some striking sequence homologies between the prokaryotic and eukaryotic enzymes, especially in regions surrounding what are probably critical cysteine residues. Conservation of these molecular properties over such a large phylogenetic range supports the argument that certain structural features are critical and must be maintained during evolution if a heme protein is to function as a P450.

The most thoroughly understood P450 is that of *Pseudomonas putida* (P450$_{CAM}$) which converts camphor to 5-*exo*-hydroxycamphor(36). The reaction cycle of P450$_{CAM}$ is shown in Scheme III where R is the substrate, camphor, and Pd is the specific reductant of P450$_{CAM}$, putidaredoxin, a two iron-two sulfur protein represented as Pd^r in the reduced form and Pd^{ox} in the oxidized form. In the first step of the reaction cycle, ferric (Fe(III)) P450 binds a camphor molecule, R in Scheme III. The second step involves reduction of the ferric enzyme-substrate complex which results in the binding of an O$_2$ molecule to ferrous P450 generating a complex resembling the

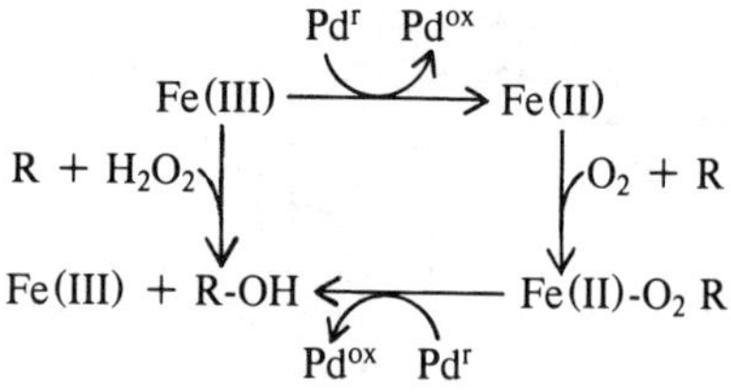

$$Fe(III) \xrightarrow{\quad Pd^r \quad Pd^{ox} \quad} Fe(II)$$

$$R + H_2O_2 \downarrow \qquad \qquad \downarrow O_2 + R$$

$$Fe(III) + R\text{-}OH \xleftarrow{\quad Pd^{ox} \quad Pd^r \quad} Fe(II)\text{-}O_2 R$$

Scheme III

oxyglobins(56, 57). Finally, a second electron is transferred from putidaredoxin resulting in cleavage of the O-O bond and hydroxylation of camphor. The second electron
transfer step is particularly interesting since efficient product formation requires the
interaction between putidaredoxin and P450. Putidaredoxin serves an effector role as
well as a source of electrons(36, 58, 59).

The precise sequence of events between the second electron transfer step
and hydroxylation of camphor remains unclear since all the intermediates have not
been identified. Nevertheless, after the second electron transfer step, O_2 is probably
reduced to the peroxy level and it is the peroxy O-O bond which is then cleaved. That
a peroxide-like intermediate may be involved is evidenced by the ability of peroxides to
substitute for O_2 and bypass the electron transfer steps as shown in Scheme III(59, 60).
By analogy with peroxidases, it seems most reasonable that the P450 peroxide O-O
bond cleaves heterolytically. However, a recent series of studies on the reaction of
microsomal P450 with various peroxides has led to the suggestion that the reaction
proceeds homolytically (generating two radicals, RO• and •OH)(61-63). However,
there are several problems with a homolytic mechanism in the physiological reaction
with O_2. First, homolysis generates destructive oxygen radicals which would be very
difficult for an enzyme to control. Second, model oxygenase systems favor a heterolytic
process(29, 64-67). Third, the molecular machinery available to enzymes, such as
acid-base catalysis and charge stabilization(68-73), are better suited to promote and
control a heterolytic mechanism.

Assuming that fission of the O-O bond proceeds heterolytically releasing
H_2O, the enzyme retains one oxygen atom with only 6 valence electrons. Possible structures of the iron-oxygen complex are Fe^{3+}-O, Fe^{4+}-O$^-$ or Fe^{5+}=O. More generally,
we can write $[Fe{-}O]^{3+}$ which ignores the two negative charges on the porphyrin core.

The hypothetical Fe-O species is termed the "oxene" intermediate(64) since the chemistry of this species should resemble that of a carbene or nitrene in that an oxene oxygen atom is an electrophilic agent capable of hydrogen abstraction and oxygen insertion (i.e. hydroxylation) reactions(64). Hydroxylation then proceeds by the enzyme rigidly orienting a substrate molecule so that the oxene oxygen atom stereospecifically combines with the substrate.

c. Oxidases

Cytochrome oxidases are far and away the most complex of the heme enzymes and since there are no crystallographic data available on oxidases, we will only briefly consider this class of heme enzyme.

Mitochondrial oxidase is a multisubunit enzyme spanning the inner mitochondrial membrane(74). A reasonably good picture is emerging of how the subunits are oriented relative to one another and to the lipid bilayer(75) as is information on the biosynthesis of the various subunits(76-79). Nevertheless, little is known about the specific function of most of the subunits.

Mitochondrial oxidase contains two a-type hemes (designated a_3 and a) and one copper atom associated with each heme. Indirect evidence suggests that the metal atoms may be associated with the largest two subunits, subunits I and II(80-82). The reaction of various ligands, including O_2, occurs at the heme a_3-Cu site while the heme a-Cu site probably functions to store electrons which are ultimately funnelled into the a_3-Cu site to complete reduction of O_2.

Several bacterial cytochrome oxidases also have been purified and characterized(83) and offer some advantages for study since these enzymes are much simpler in subunit composition. *Paraccocus denitrificans* cytochrome oxidase is a particularly good example. This oxidase consists of two subunits containing two hemes and two coppers. The two subunits are similar in size to mitochondrial subunits I and II and subunit II from the bacterial and mictochondrial oxidases cross react immunologically(83). This latter similarity is an additional reason for suspecting that the metal atoms are associated with subunits I and/or II of mitochondrial enzymes.

One primary objective in studying cytochrome oxidase is to understand how the 4-electron reduction of O_2 proceeds. *A priori* it is reasonable to assume that the reaction occurs in steps requiring the formation of stable reduced intermediates of O_2.

In fact, such intermediates have been trapped at low temperatures and one of these is thought to be a peroxy intermediate complexed with the heme a_3-Cu site(84). Relevant to these studies are the recent findings that H_2O_2 reacts with cytochrome oxidase to yield a spectrally distinct product(85) and that H_2O_2 can substitute for O_2 in the cytochrome oxidase catalyzed oxidation of cytochrome c(85, 86). This ability to use H_2O_2 rather than O_2 is similar to what was found in P450. Therefore, all three types of heme enzymes, peroxidases, oxygenases, and oxidases, have an Fe-peroxy complex at some point in their respective reaction cycles and each ultimately operates as a peroxidase by cleaving the peroxy O-O bond. One might look for mechanistic similarities shared by all three enzymes such as heterolytic fission of the peroxy intermediate and the formation of heme or protein centered free radicals.

III. HEME ENZYME STRUCTURES

a. Accuracy of Protein Structures

Before discussing specific crystal structures, a few brief words on the accuracy of protein structures are warranted. With few exceptions, protein structures are determined using the method of multiple isomorphous replacement (MIR). Interpretation of MIR electron density maps is a much more subjective exercise than is often appreciated and a considerable amount of common sense and intuition go into this phase of a structure determination. In most cases, the course of the polypeptide chain can be determined but the precise location of all sidechains may be difficult. Even though initial MIR structures reveal a wealth of useful information, it has become apparent that many of the more subtle features which cannot be seen in MIR maps are just those fine details that may be important to the protein's function. Fortunately, in recent years, with advances in computer technology and software, protein crystal structure refinement has become possible, enabling a much more accurate determination of three dimensional crystal structures.

Briefly, crystal structure refinement proceeds as follows. Theoretical structure factors (F_cs) are computed from initial estimates of atomic positional parameters and compared to the observed structure factor amplitudes (F_os). The problem then becomes one of altering the positional parameters in an iterative process so as to minimize the expression $R = (\Sigma |F_o - F_c|) / (\Sigma F_o)$. This is not an altogether straightforward process because of the large number of atomic positions and structure factors involved and the unforseen pitfalls that accompany such an undertaking. The important point, however, is that refined structures are considerably more accurate. In addition, the value of R can be used to assess the accuracy of the positional parameters. For a 2.0 Å structure with R = 0.20, the approximate error in determining the position of the atoms is ±0.2 Å. This provides a quantitative estimate not available in MIR structures by which to gauge our confidence in drawing important structure-function conclusions based on x-ray structures.

b. Cytochrome c Peroxidase Structure

The 1.7 Å structure of CCP has been refined in our laboratory to R = 0.21. During the refinement, one notable error was detected in the chemically determined sequence. One of us (B.F.) found that an Asn (or Asp) must be inserted between residues 163 and 164. An Asn was preferred owing to hydrogen bonding interactions with the sidechain of Gln 256 of a neighboring CCP molecule in the unit cell. Independently of our work, Kaput et al. (87) sequenced the CCP gene and found the same change, confirming that residue 164 is Asn. Therefore, CCP consists of 294 residues and the new sequence numbering will be used in this article.

An alpha carbon backbone model of the refined structure of CCP is shown in Fig. 1. 46% of the residues are in a right handed α helical conformation, 6.1% are in a 3_{10} helical conformation, 12.4% are involved in antiparallel β pairs, and there are no parallel β or extended sheet structures. This predominance of helical secondary structure is a common feature shared by many heme proteins. CCP also contains one uncommon region of secondary structure. Residues 36-40 form one turn of left handed α-helix on the surface of the protein.

As with other enzymes of similar size, CCP folds into two clearly defined domains with the active site situated at the domain interface. The heme is sandwiched between two antiparallel helices, one from domain I (residues 42 to 54) and one from domain II (residues 164 to 171). This antiparallel arrangement of helices is also found in the globins(88), b cytochromes(89, 90), and cytochrome c'(91) though these smaller heme proteins lack the multiple domain composition.

Interactions between the heme and protein in CCP are very similar to those found in other heme proteins in that the heme is surrounded by hydrophobic residues. Two noticeable differences, however, are the degree of heme exposure and the interac-

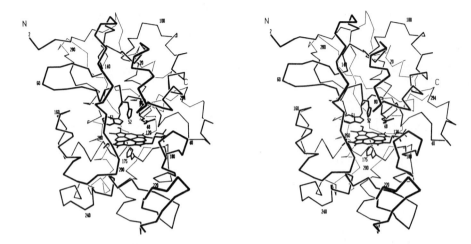

FIG 1.   Stereoscopic alpha carbon backbone model of the 1.7 Å refined cytochrome *c* peroxidase structure.   Key side chains are labeled.   The proximal ligand is His 175 while the distal ligation pocket is formed by Arg 48, Trp 51, and His 52.

tions between the heme propionates and the protein.   In myoglobin and cytochrome b_5, the heme edge is exposed and the heme propionates must bend back to form favorable hydrogen bonds on the surface of the protein.   In CCP, the heme is further from the molecular surface with no edge of the heme directly accessible, and even though the propionates are fully extended, they are prevented from directly contacting the surrounding external medium by a rigid network of hydrogen bonds (Fig. 2).   The propionate of pyrrole IV is closest to the surface but is completely sequestered behind the helical turn of residues 181-185 which reverses the direction of the polypeptide chain to form a short section of antiparallel β structure (not shown in Fig. 2).   The propionate of pyrrole IV forms hydrogen bonds with the sidechains of His 181, Asn 184, Ser 185, and a buried water molecule.   The propionate of pyrrole III lies at the bottom of a small recess in the molecular surface and, with the exception of a hydrogen bond with the amide nitrogen of Lys 179, is completely hydrated.

His 181 resides in a rather unusual environment in that the imidazole ring is bracketed by two carboxylate groups (Asp 37 and a heme propionate, Fig. 2).   As a result, we might expect His 181 to exhibit some unusual chemical properties, such as an elevated pK.   In fact, we have found that two classes of histidines in CCP can be

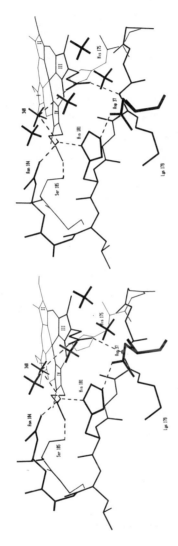

FIG. 2. Stereoscopic representation of the CCP heme propionate environment. Some side chains have been omitted for clarity. Note the hydrogen bonding between pyrrole IV propionate and residues 181, 184, and 185. Also note that His 181 hydrogen bonds with Asp 37 as well as pyrrole IV propionate. The tetrahedra in this and all figures represent ordered water molecules.

distinguished by reaction with the histidine specific reagent diethylpyrocarbonate(92). One class reacts at pH 7 with no resulting loss in enzymic activity while a second class reacts only at pH > 7 resulting in more than a 95% loss of enzymic activity(93). We suspect that His 181 belongs to this latter group of high pH histidines. Its interactions with two carboxylates should increase the pK of the imidazole group thereby rendering His 181 unreactive as a nucleophile at neutral pH.

As in ferric myoglobin and hemoglobin, the axial ligands in CCP are a histidine residue (His 175) and a water molecule (see Fig. 3). At the pH of the crystallization medium (6-6.5), CCP is in the high spin ferric state(94) so we can conclude that high spin ferric CCP is hexacoordinate.

In addition to coordinating with the iron atom, the proximal histidine ligand can have significant electronic effects on the heme and its substituents (methyls, vinyls, and propionates) depending on the orientation of the imidazole ring about the proximal N-Fe bond. NMR techniques have been used to study such effects and distinct differences have been found in the hyperfine resonances of the heme methyl protons in CCP and myoglobin(95). The heme methyl pattern in CCP is $5 > 1 > 8 > 3$ with methyl 5 exhibiting the furthest down field shift and 3 the least, while in myoglobin the pattern is $8 > 5 > 3 > 1$(95). These differences are readily rationalized on the basis of the x-ray structures. As shown in Fig. 4, the orientation of the imidazole ring about the N-Fe bond is about the same in both proteins, but in CCP the heme is flipped 180° relative to myoglobin. As a result, the histidine contributes greater unpaired spin density to methyls 5 and 1 in CCP resulting in a down field shift for these protons. In myoglobin, the effect is just the opposite.

At least part of the reason for the observed orientation of the proximal histidine in CCP is the hydrogen bonding network involving His 175 shown in Fig. 3. ND1 hydrogen bonds with an internal aspartate (Asp 235) while Asp 235 forms a second hydrogen bond with the indole ring nitrogen of Trp 191 and a third with an internal water molecule. As a basis for comparison, in all other heme proteins for which x-ray data are available, with the exception of cytochrome c'(91), the proximal histidine hydrogen bonds with a carbonyl oxygen atom(96). A carboxylate group should have a higher affinity for the histidine ND1 proton so we might expect the Asp-His hydrogen bond in peroxidase to be stronger than the carbonyl oxygen-His hydrogen bond in the globins. NMR data(97) indicate that the proximal imidazole in peroxidases is indeed more strongly hydrogen bonded.

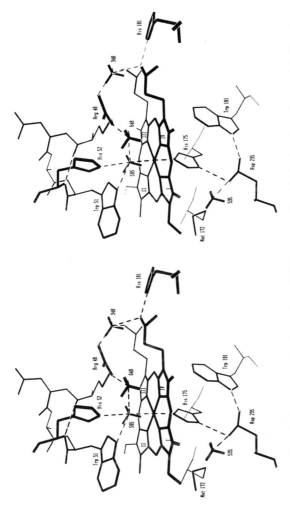

FIG. 3. Detailed stereoscopic view of the CCP heme crevice displaying the substrate binding site on the distal side of the heme (top) and the proximal histidine environment (bottom). Hydrogen bonds referred to in the text are indicated with dashed lines.

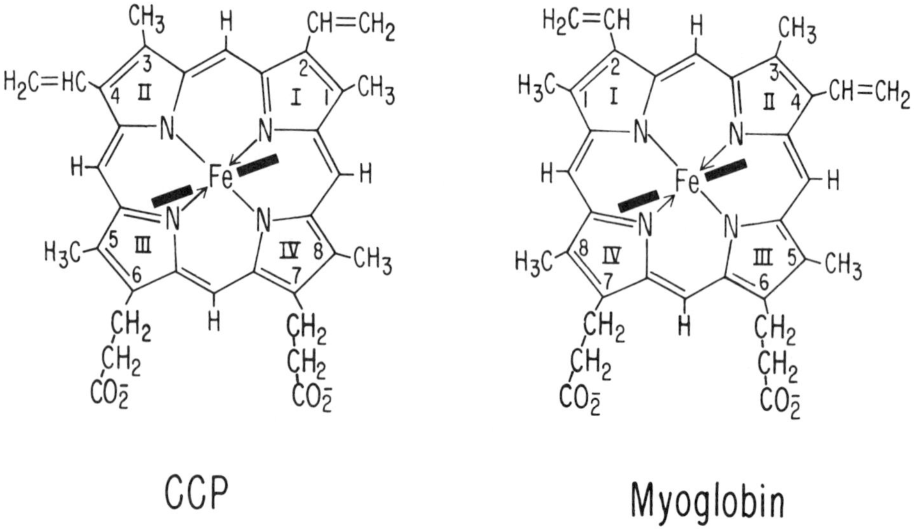

FIG. 4. Orientation of the proximal histidine in CCP and myoglobin. Both views are along the proximal N-Fe bond looking down on the proximal surface. This figure is based on a similar figure in reference (95).

On the distal side of the heme, Arg 48, Trp 51 and His 52 form a ligation pocket for peroxides and other ligands (Fig. 3). These three residues form one turn of α-helix with the carbonyl oxygen atom of Arg 48 hydrogen bonding with the amido nitrogen atom of His 52. Arg 48 lies near the entrance to the crevice while Trp 51 is the most deeply embedded residue with its indole ring nearly parallel to and contacting the heme. An additional important structural feature on the distal side that became visible only as a result of crystallographic refinement is the ordered water structure. As shown in Fig. 3, the axial water ligand (Wat 595) hydrogen bonds with His 52 and the indole nitrogen of Trp 51. A second water molecule (Wat 648) interacts with the axial water ligand while a third (Wat 348) is situated between the guanidinium group of Arg 48 and both propionates.

The distal ligation pocket is accessible to solvent and small molecules owing to a channel connecting the distal side of the heme crevice with the protein surface (Fig. 5) and is the only obvious route for ligands including peroxides to and from the active site. The edge of pyrrole IV lies at the bottom of this channel about 10 Å from

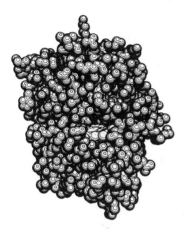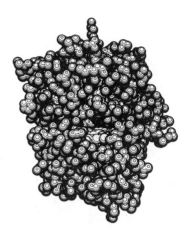

FIG. 5.   Stereoscopic space filled model of CCP looking into the heme access channel. Protein atoms are dark while heme atoms are lightly shaded. The edge of pyrrole IV is visible and is situated about 10 Å from the opening of the access channel. This figure was generated by the program BALSTK written by one of us (B.F.).

the molecular surface. As a result of the ligand access channel, Trp 51 is the only distal side residue among the three we have considered which is truly buried.

To what degree is the active site structure just described conserved in plant peroxidases? A number of lines of evidence, including comparative sequence data(1-5), indicate that the proximal ligand is also a histidine residue. Table 1 shows that sequences around the distal and proximal histidines are very similar. Of particular importance is the conservation of Arg 48, His 52, and an aromatic residue at position 51 on the distal side.

c. Catalase Structure

The 2.5 Å crystal structure of beef liver catalase has been determined in Michael Rossmann's laboratory by Murthy *et al.*(98). Though the structure has not been refined, initial MIR maps were considerably improved by averaging the electron density of subunits related by non-crystallographic symmetry(98). As a result, an accurate tracing of the subunit polypeptide backbone and location of most sidechains was possible.

What is most striking about the catalase structure is how different it is from the peroxidases. Catalase is a tetramer while the plant peroxidases and CCP are mono-

TABLE 1

Amino Acid Sequences Around the Distal and Proximal Histidines in Cytochrome
c Peroxidase (CCP, ref. 4), Horseradish Peroxidase c (HRPc, ref. 3) and Various
Isozymes of Turnip Peroxidase (P_1, P_2, P_3 and P_7, ref. 1 and 2). Numbering is Based
on the CCP Sequence. The Distal and Proximal Histidines are Underlined.

Distal Side

				45					50					55					60
CCP	Y	G	P	V	L	V	R	L	A	W	<u>H</u>	T	S	G	T	W	D	K	H
HRP<u>c</u>	I	A	A	S	I	L	R	L	H	F	<u>H</u>	D	C	F	V	N	G	C	D
P_1	I	G	A	S	L	I	R	L	H	F	<u>H</u>	D	C	F	V	N	G	C	D
P_2	I	G	A	S	L	I	R	L	H	F	<u>H</u>	D	C	F	V	K	G	C	D
P_3	I	G	A	S	L	I	R	L	H	F	<u>H</u>	D	C	F	V	N	G	C	D
P_7	M	G	A	S	L	I	R	L	F	F	<u>H</u>	D	C	F	V	N	G	C	D

Proximal Side

CCP	V	A	L	M	G	A	<u>H</u>	A	L	G	K	T	H
HRP<u>c</u>	V	A	L	S	G	G	<u>H</u>	T	F	G	K	N	Q
P_1	V	V	L	S	G	A	<u>H</u>	T	F	G	R		
P_2	V	V	L	S	G	A	<u>H</u>	T	F	G	R		
P_3	V	V	L	S	G	A	<u>H</u>	T	F	G	R		
P_7	V	V	L	S	G	A	<u>H</u>	T	I	G	R	S	R

mers. The considerably larger catalase subunit is divided into four domains and has a
relatively low helical content (26%). In addition, the proximal heme ligand in catalase
is not histidine but tyrosine.

Despite these differences, certain features shared by CCP and catalase set
these proteins apart from the globins and cytochromes. The heme in both enzymes is
not exposed at the surface and, in fact, the heme in catalase is even further removed
from the surface than in CCP. Furthermore, catalase also has a channel connecting the
distal side of the heme with the surface(98).

d. Preliminary Crystallographic Results on Cytochrome P450

Cytochrome $P450_{CAM}$ crystallizes in a form suitable for high resolution x-ray
diffraction studies(99). We are currently in the process of interpreting a 2.8 Å MIR

map and do not as yet have a detailed molecular model. Nevertheless, a cursory examination of the electron density map reveals the overall size and shape of P450, location
of the heme and the predominance of helical structure (Figs 6 and 7). As shown in
Figs 6 and 7, the heme appears as a flat disk of density with the heme normal almost
perfectly aligned with the crystallographic Z axis. One surface of the heme is close to
and faces toward the surface of the enzyme but is nevertheless masked from exposure
to solvent by the protein. Especially noteworthy is a helical segment running along the
left side of the heme in Fig. 7. Strong density connecting this helix with the heme iron
atom (see arrow, Fig. 7) indicates that the presumed cysteine ligand is part of this
helix. We assume, therefore, that the P450 substrates, camphor and O_2, bind on the
opposite or interior side of the heme (right side of heme, Fig. 7). In fact, an isolated
lobe of density just above the interior surface of the heme is about the right size and
shape for a camphor molecule though we will have to confirm this by difference Fourier
analysis.

From these preliminary observations, it is apparent that the organization of
the heme and its ligands is different in P450 and CCP. In CCP the proximal ligand is
situated in an internal pocket while this appears not to be the case for P450. Despite
this important difference, P450 more closely resembles CCP and catalase than the cytochromes in that no edge of the heme is directly accessible at the surface of the protein.

e. Active Site Comparisons

Fig. 8 shows the active site structures of CCP, catalase and myoglobin.
First, consider the structures on the distal (upper) side of the heme. All three proteins
have a histidine residue on the distal side, though the precise location is different in
each case. In myoglobin and CCP the distal histidine is nearly perpendicular to the
heme and hydrogen bonds with the axial water ligand though in CCP the distal histidine
is further from the heme. In catalase the distal side histidine (His 74) is situated quite
differently since the imidazole ring is nearly parallel to the heme and occupies approximately the same position as Arg 48 in CCP. In myoglobin there is no homologue to
Arg 48. In catalase a valine (Val 73) is located near the position of the distal histidine
in CCP and myoglobin. Where myoglobin has a phenylalanine (Phe 43), catalase has
an asparagine (Asn 147), and where both CCP and catalase have an aromatic residue
parallel to and contacting the heme (Trp 51 in CCP, Phe 160 in catalase), myoglobin
has a valine (Val 68). Referring back to Table 1, note that a phenylalanine is found at

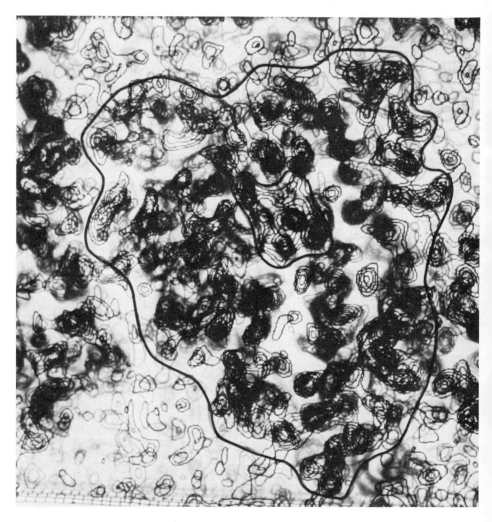

FIG. 6.   Approximately 12 Å thick section of the 3.0 Å P450$_{CAM}$ electron density map viewed along the crystallographic Z axis. A single P450 molecule and the heme are outlined.

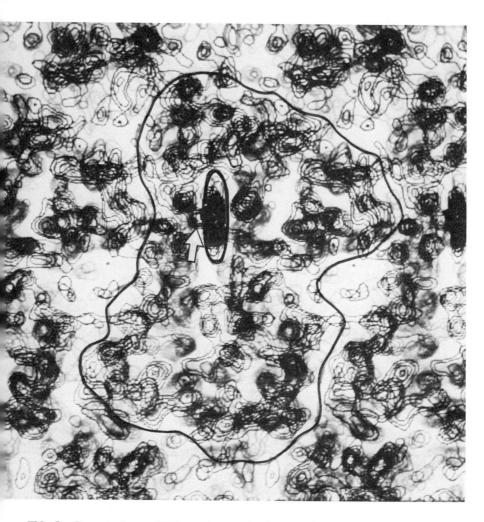

FIG. 7. The cytochrome P450$_{CAM}$ electron density map viewed along the crystallo-graphic Y axis. Density corresponding to the cysteine ligand is indicated with an arrow. Locations of the camphor and O_2 binding sites referred to in the text are on the *right* side of the heme opposite the proximal cysteine.

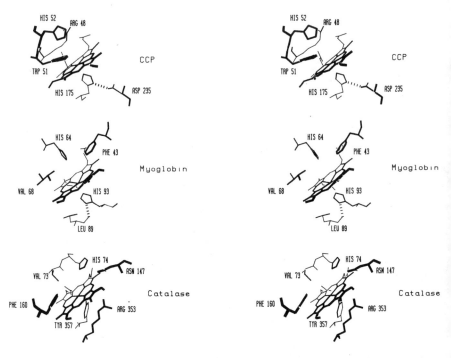

FIG. 8.   Comparisons of the active site structures in CCP (top), myoglobin (center), and catalase (bottom). Coordinates for myoglobin were obtained from the Brookhaven Protein Data Bank(100) and are the refined 2.0 Å coordinates of Takano(101). Catalase coordinates from the 2.5 Å structure were generously supplied by Dr. Michael Rossmann. The dashed lines represent hydrogen bonds between ND1 of the proximal histidine and Asp 235 (CCP) or the carbonyl oxygen atom of Leu 89 (myoglobin).

the position corresponding to Trp 51 of CCP in all plant peroxidases for which sequence data is available, emphasizing the importance of an aromatic residue at this location in both peroxidases and catalase.

From the preceding comparisons, it is clear that CCP and catalase have two residues adjacent to the ligand binding site (Arg 48, His 52 in CCP and His 74, Asn 147 in catalase) capable of participating in acid-base catalysis and charge stabilization required in the formation of Compound I. Not too surprisingly, model studies(102) have shown that the more polarizing environment near the peroxide binding site found in the peroxidases and catalase will promote cleavage of the O-O bond. Therefore, one reason why myoglobin is a poor peroxidase is the lack of suitably positioned residues capable of polarizing the O-O bond. Instead, the more hydrophobic globin distal environment is conducive to reversible O_2 binding. On the other hand, the more polarizing environment in the heme enzymes is not compatible with reversible O_2 binding since polarization of the O-O bond will promote iron oxidation.

An especially noteworthy difference between CCP and catalase, which we wish to emphasize, is the location of His 74 in catalase compared to His 52 in CCP, and the absence of Arg 48 (conserved in all peroxidases (Table 1)) but the presence of an asparagine residue in catalase. Such contrasting structural features undoubtedly reflect the different catalytic requirements of catalase versus peroxidase. It would appear that Nature has solved the problem of O-O bond heterolysis by providing a suitable polarizing environment in heme enzymes but achieves the required environment in a variety of ways.

Next, consider the structures on the proximal side of the heme. The most obvious difference is that histidine serves as the proximal ligand in the peroxidases and globins while this residue is tyrosine in catalase (98) and cysteine in P450(37-50). A more subtle difference described in the last section is the hydrogen bonding interactions between the proximal histidine and other groups. The stronger His-carboxylate hydrogen bond in CCP vs. the carbonyl oxygen-His interaction in the globins should impart a greater anionic character to the peroxidase proximal histidine. Because the more anionic proximal histidine in CCP is better able to stabilize the Fe^{3+} state, CCP has a lower midpoint potential of the Fe^{3+}/Fe^{2+} couple (-194 mV)(103) than does myoglobin (+50 mV)(104). In catalase the situation may be similar since the proximal ligand is thought to be the anionic form of Tyr 357(98).

The potential importance of the Asp 235-His 175 interaction in CCP catalysis is evidenced by the pH dependence of the reaction between CCP and H_2O_2 to give Compound I. In CCP, a group with a pK = 5.5 must be unprotonated to realize a maximum rate of Compound I formation(105). The crystal structure suggests two possible candidates. As we will discuss shortly, His 52 must be unprotonated to serve as an effective acid-base catalyst in the reaction with peroxides, suggesting the possibility that the pK = 5.5 group might be His 52. A pK = 5.5 is somewhat low for histidine, but active site residues often exhibit unusual properties. Nevertheless, His 52 is readily accessible to solvent molecules via the ligand access channel and we see no compelling reason, based on the crystal structure, why His 52 should have an unusual pK. A

second and more likely candidate for the pK = 5.5 group is Asp 235. Protonation of
Asp 235 will weaken the Asp 235-His 175 hydrogen bond and thereby disrupt the intri-
cate proximal side hydrogen bonding network (Fig. 6). As a result, His 175 will no
longer be able to effectively stabilize higher oxidation states of the iron during the formation of Compound I and the reaction rate will slow. A pK = 5.5 is unusually high
for aspartic acid, but unlike His 52, Asp 235 does reside in an unusual environment
that could elevate its pK since the Asp 235 sidechain is buried in the internal "proximal
pocket". As a basis for comparison, studies with lysozyme have shown that the interior
active site carboxylates have pKs > 5 (106) and the active site internal Asp 26 of dihydrofolate reductase has a pK > 7(107).

If Asp 235 plays a critical role in peroxidase catalysis, one might expect a
similar set of interactions to be found in other peroxidases. Examination of the known
peroxidase sequences in the vicinity of Asp 235 reveals some interesting similarities and
differences.

		235							
CCP	T	$\underline{D}$	Y	-	S	L	I	Q	D
HRP$_c$	E	$\underline{E}$	Q	K	G	L	I	Q	S
P$_7$	A	A	Q	R	G	L	I	H	H

Similarities in the sequences shown above could mean that Glu 239 in HRP$_c$
(underlined above) is analogous to Asp 235 in CCP and interacts with the proximal histidine. The absence of an acidic residue at this position in the P$_7$ isozyme of turnip
peroxidase could be one reason why P$_7$ has a higher midpoint potential and is more
easily reduced than other peroxidases (-166 mV for P$_7$ (108) versus -194 mV for CCP
(105) and less than -200 mV for other plant peroxidases(108)).

IV. MECHANISM OF COMPOUND I FORMATION

The second order rate constant for the reaction between H_2O_2 and peroxi-
dases is 10^7 to 10^8 $M^{-1}sec^{-1}$(109, 110) to be compared with 10^{-3} $M^{-1}sec^{-1}$ for the reaction between heme and peroxides at physiological pH(111). In order to account for this
remarkable rate enhancement, the enzyme must function at least in part by stabilizing
the energetically unfavorable charge separation brought about in forming species resembling RO^- and ^+OH from the RO1-O2H(112) substrate in the activated complex. Such
stabilization is best achieved by acid-base catalysis and stabilization of the developing

negative charge on $RO1^-$(5, 113, 114). The rigid orientation of the substrate required to achieve the necessary stabilization is suggested by the large negative entropy of activation in forming Compound I(115).

With these ideas in mind, we studied the interaction of a hypothetical RO1-O2H peroxide substrate with CCP given the constraints that O2 coordinates with the iron atom and that an alkyl R group must be accommodated in the active site(5). We find that the substrate O2 atom replaces the axial water ligand (water 595) while the O1 atom replaces water 648 (see Fig. 4). Our initial model building studies were carried out before the active site water molecules were located, but even so, it was clear that the substrate must occupy these positions. Apparently waters 595 and 648 form a "ghost" of the substrate.

Figure 9 outlines the proposed mechanism of Compound I formation. Upon entering the active site, the substrate O2 atom coordinates with the iron atom with concomitant removal of the O2 proton by His 52. Deprotonation of the peroxide molecule by His 52 need not present a serious thermodynamic barrier even though the pK of $H_2O_2 \simeq 12$ because coordination of the O2 atom with the heme iron will significantly weaken the O2-H bond. His 52 completes its role as an acid-base catalyst by transferring the O2 proton to O1.

That the substrate remain protonated upon entering the active site is critical to the acid-base role of His 52. Therefore, it is important to note that the peroxidase active site accepts only protonated ligands (HF, HCN) and protonated substrates (115-121) while the globins accept anions(122). The reason why peroxidases do not bind anions is a longstanding problem but a comparative structural analysis between the globins and CCP suggests that the difference in ligand binding properties is related to the preferred hydrogen bonding interactions between the distal histidine and ligands. Two different types of hydrogen bonding interactions are possible since there are two tautomeric forms of the imidazole ring at physiological pH.

In both the globins and CCP, N2 hydrogen bonds with the axial water ligand. In the

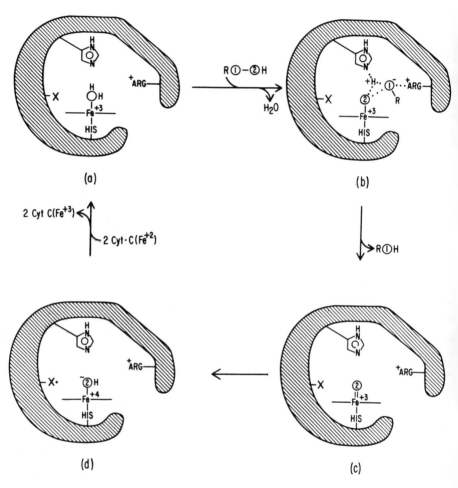

FIG. 9. Mechanism of CCP Compound I formation. a) the native enzyme; b) the activated complex with the distal histidine operating as an acid-base catalyst and the active site arginine stabilizing a developing negative charge on RO1; c) the hypothetical oxene intermediate and; d) Compound I after the intramolecular electronic rearrangement of (c) to give Fe(IV) and a free radical, X•.

globins, N1 is exposed at the molecular surface and is readily available for hydrogen bonding with solvent molecules, either as an acceptor (tautomer I) or donor (tautomer II). There appears to be no structural basis for preferring either tautomer in the globins so N2 can either donate a hydrogen bond to an X^- ligand or accept a hydrogen bond from an HX ligand. In CCP, however, N1 forms a 2.7 Å hydrogen bond with the sidechain carbonyl oxygen atom of Asn 82. We are certain that the Asn sidechain oxygen and not nitrogen hydrogen bonds with the distal histidine owing to a clearly defined hydrogen bonding interaction between the Asn 82 sidechain nitrogen atom and a peptide carbonyl oxygen atom. Therefore, tautomer II is favored in CCP meaning that N1 can serve only as a hydrogen bond acceptor. As a result, ligands capable of donating hydrogen bonds to N2 (HX but not X^-) form the most thermodynamically stable complexes and are preferentially bound to CCP. Most importantly, this hydrogen bonding pattern insures that CCP accepts only protonated substrates.

Arg 48 further promotes heterolysis by stabilizing a developing negative charge on O1. As shown in Fig. 3, the guanidinium group of Arg 48 is only 2.5 Å from water 648 (i.e. substrate O1 atom) so we can expect an excellent hydrogen bonding interaction.

To briefly summarize, the exceptional rate of Compound I formation can be attributed to at least 4 factors: 1) coordination of the O2 peroxide atom with the ferric iron atom significantly weakens the O2-H bond; 2) directed proton transfer or acid-base catalysis by His 52; 3) charge stabilization by Arg 48, and 4) stabilization of higher iron oxidation states by the proximal histidine.

Heterolysis of the O-O bond leaves the hypothetical [Fe-O] intermediate which we described earlier in the P450 section. This species is highly electrophilic and readily undergoes an intramolecular electronic rearrangement resulting in a ferryl (Fe^{+4}) iron and an organic radical (Fig. 9). There is now sufficient evidence to indicate that in catalase and plant peroxidases X• is a cationic heme radical(123-126). However, in CCP the radical is centered on an amino acid sidechain(127). The precise identity of this sidechain has yet to be unambiguously established.

Initially, we suggested that the CCP radical was centered on Trp 51 for the following reasons. First, the EPR properties of CCP Compound I suggested an aromatic radical(127). Second, the spontaneous decay of CCP Compound I without exogenous reductants results in internal redox reactions leading to the destruction of tryptophan(128, 129). Third, an ultraviolet difference spectrum is generated upon formation of CCP Compound I indicating perturbation of aromatic residues near the site of peroxide heterolysis(130). Fourth, Trp 51 is directly adjacent to and must interact with the peroxide O2 atom (Fig. 3). Therefore, Trp 51 is well positioned to give up a hydrogen atom to O2 generating an indole free radical. In catalase and the plant peroxidases, Trp 51 is replaced by Phe (Table 1) and because Phe is less readily oxidized than Trp, the heme is preferentially oxidized. Schematically one can view this as a simple equilibrium

$$X \bullet \text{ heme (red)} \longleftarrow \; \longrightarrow X \text{ heme} \bullet \text{ (green)}$$

where the radical equilibrates between the aromatic residue, X, and the heme. In CCP the equilibrium lies to the left while in plant peroxidases it lies to the right. The above equilibrium accounts for the different spectral properties of CCP and plant peroxidase Compound I since cationic heme radicals are green(124) while the ferryl-oxo species with no heme radical is the same color as Compound II, red.

Unfortunately, the situation is not quite so simple. Hoffmann et al. (131) have carried out detailed EPR and ENDOR studies on CCP Compound I and concluded that an aromatic radical is unlikely. Comparisons with model compounds lead these authors to suggest that a sulfur centered radical is more reasonable. The most logical candidate is Met 172 (Fig. 3). The Met 172 sulfur atom is $\sim$4.1 Å below the heme, so one could envision an electron transfer equilibration between the two. This equilibration is impossible in the plant peroxidases where, by sequence comparison, position 172 is occupied by a serine residue (see Table 1). One important requirement for a Met centered free radical is the need for stabilization of the cationic sulfur radical by a suitable nucleophile(131). Asp 235 is close enough (4.3 Å) to Met 172 for this purpose.

To further complicate matters, Fujita et al. (132) demonstrated that in model systems significant transfer of charge density from the heme to the proximal ligand can occur, implying that the radical might be delocalized over the proximal histidine. In CCP, Trp 191 is stacked immediately adjacent to His 175 (Fig. 3) offering an extended π system for delocalization of the radical on the proximal side of the heme. Unfortunately, the crystal structure does not offer a simple choice between these possibilities.

V. HEME ENZYME ELECTRON TRANSFER REACTIONS

a. Introduction

Electron transfer between macromolecular redox pairs presents two fundamental problems. One is the distance between redox pairs and whether or not protein groups participate in the electron transfer reaction, especially when relatively large electron transfer distances are involved. Second is the problem of specificity. How do the redox partners recognize and bind to one another?

To approach the question of distance we first categorize biological electron transfer reactions into two basic types, those involving short distances (< 10 Å) and those involving long distances (> 10 Å). In short distance electron transfer reactions, both the donor and the acceptor heme edges are accessible at the surface of the protein while in long distance reactions, at least one heme is not directly accessible (Fig. 10). We suggest that in short distance electron transfer reactions neither the donor nor the acceptor molecules are enzymes but electron carriers like the cytochromes. In long distance reactions, the acceptor is an enzyme or a high energy oxidant like the chlorophyll radical. The available estimates for distances between redox centers support this suggestion. Results obtained using various spectral probes indicate that the distance between redox centers in cytochrome c and acceptors like cytochrome oxidase(133) and bacteriochlorophyll(134) are long (> 20 Å). Similar experiments with the CCP-cytochrome c redox pair give a 14 Å estimate(135). Furthermore, the crystallographic data demonstrates that in the three heme enzyme examples described earlier, (CCP, catalase, and P450) the heme edge is not directly accessible while in all known cytochrome structures it is accessible.

The reason for the differences in heme exposure and, therefore, electron transfer distances is understandable when we consider the characteristic functional difference between heme enzymes and cytochromes. In heme enzymes, reactions with O_2 and peroxides generate oxygenated iron and heme or protein centered radicals as intermediates while no similar intermediates are formed in the cytochromes. Such intermediates are highly reactive oxidizing agents and must be sequestered within the insulating environment of the polypeptide and/or membrane. This protects the oxidizing center from nonspecifically discharging owing to random encounters with cellular

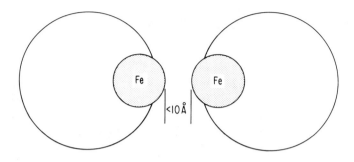

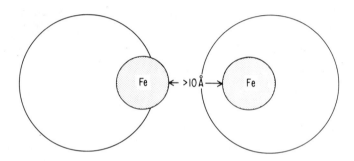

FIG. 10. Schematic representation of short and long distance electron transfer complexes. In the short distance complexes (top), the hemes of both donor and acceptor proteins are accessible at the molecular surface. In long distance complexes (bottom), the acceptor heme is sequestered within the folds of the polypeptide chain and is not directly accessible.

oxidants and reductants. Moreover, since the heme is not directly accessible, protein groups may serve as a path for electron transfer, a topic we will consider in part (d) of this section.

b. Specificity

Before considering how electron transfer occurs in both long and short distance reactions, we must first turn to the question of how two redox partners recognize one another to form a prereaction intermolecular complex. Specificity is best understood for those reactions involving cytochrome c. Chemical modification(136-148) and crystallographic data(149-151) have demonstrated that the highly conserved lysines surrounding the exposed edge of the cytochrome c heme are required for the recognition and binding to its electron transfer partners (Fig. 11). These partners include both oxidants, like cytochrome c oxidase(137-139, 147) CCP(142-144) and sulfite oxidase(148) and reductants like cytochrome c reductase(140-142, 147) and cytochrome b_5(147). The prevailing view emerging from these studies is that positive charges lining the exposed edge of the cytochrome c heme recognize and bind to a set of complementary negative charges on both cytochrome c reductases and oxidases, allowing oxidation and reduction to occur via the exposed heme edge(149-152).

Specificity of interaction is also evident from the kinetics of cytochrome c electron transfer reactions. The second order rate constant for the reaction between CCP and cytochrome c, $\sim 10^8 \text{M}^{-1}\text{sec}^{-1}$(153), approaches the limits imposed by the diffusion of both proteins indicating that nearly every encounter between CCP and cytochrome c results in a productive complex. Therefore, CCP and cytochrome c must be correctly oriented at a relatively long distance from one another owing to proper

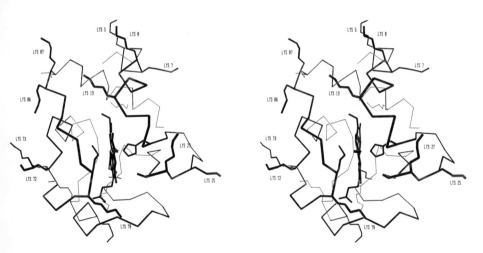

FIG. 11. Stereoscopic view of cytochrome c viewed along the exposed heme edge. All lysines are shown and those lysines lining the exposed heme edge are labeled. The surface closest to the viewer is the cytochrome c recognition domain to which both macromolecular oxidants and reductants of cytochrome c bind.

orientation of complimentary electrostatic fields and molecular dipoles(154). In fact, Koppenol and Margoliash have shown that the rate of electron transfer is indeed sensitive to changes in the cytochrome c dipole(154).

On the other hand, the first order rate of electron transfer or turnover number for biological electron transfer reactions is not remarkably different than non-physiological reactions. For example, the turnover number for cytochrome oxidase is $300 \, sec^{-1}$(155) while that for CCP is $380 \, sec^{-1}$(156). Compare these rates with $20 \, sec^{-1}$ for the transfer of electrons between the heme iron atom of cytochrome c and a ruthenium atom covalently attached to the surface of cytochrome c 15 Å from the heme iron atom(157). While the enzymic rates are about 15 to 40 times faster than the non-physiological example, the rate increase does not nearly approach the enhancements often observed in nonelectron transport enzymes where turnover numbers are typically several orders of magnitude larger than nonenzymic rates. Kraut(158) has suggested that electron transfer reactions are in fact intrinsically fast and that the primary role of the heme enzyme is not to accelerate the rate of electron transfer but rather to insure specificity by pairing with the correct redox partner. As outlined above, the required specificity is achieved by sequestering the heme within the insulating environment of the polypeptide chain in order to prevent random discharge of reactive intermediates, by providing a complimentary set of electrostatic interactions to insure correct orientation and pairing, and by providing a path for electron transfer once the correct complex is formed.

c. Hypothetical Electron Transfer Complexes

While the cytochrome c interaction "site" has been accurately mapped, the exact mode of interaction between donor and acceptor hemes remains to be established. Some indication of these interactions is suggested by four hypothetical electron transfer complexes involving cytochrome c. Salemme was the first to propose such a model for the cytochrome c-cytochrome b_5 complex(150) which is shown in Fig. 12. Though cytochromes c and b_5 are not physiological redox partners, b_5 does reduce cytochrome c at a substantial rate *in vitro*(159) and formation of a 1:1 complex has been demonstrated(160). Furthermore, the close similarity between amino acid sequences of b_5 and the heme binding domain of yeast cytochrome b_2(161) suggests that the b_5-c model may, in fact, be relevant to the *in vivo* transfer of electrons from cytochrome b_2 to cytochrome c.

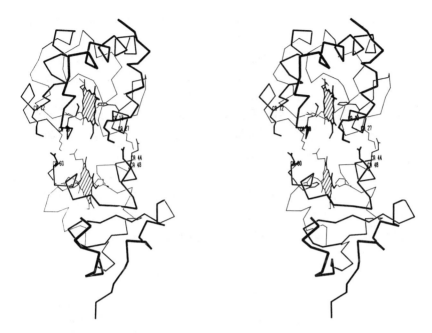

FIG. 12. Stereoscopic view of the hypothetical cytochrome b_5 (bottom) - cytochrome c (top) complex. Residues involved in intermolecular ionic contacts are labeled. The model shown was produced with a computer graphics system and is essentially the same model described by Salemme(150).

In the b_5-c model, the essential and highly conserved lysines are seen to interact with a set of carboxylates lining the exposed edge of the b_5 heme (see Fig. 12). Subsequent chemical modification studies of b_5 confirmed the prediction of which carboxylates are involved in the b_5-c complex(162). In addition to the resonably good fit between oppositely charged surfaces, the key feature of the b_5-c model is that the hemes are coplanar with an edge to edge distance of 8 Å thereby allowing direct electron transfer between hemes.

A second model which has physiological importance is the cytochrome b_5-methemoglobin complex(163). In the methemoglobin reductase system, b_5 serves as the electron donor to methemoglobin (164, 165). The direct interaction between b_5 and methemoglobin, as well as the importance of electrostatic forces in forming the complex, has been established(166). As a result, one might expect the b_5-methemoglobin complex to resemble the b_5-c complex. Indeed, the distribution of lysines surrounding the exposed edge of the hemoglobin heme in both the α and β chains is remarkably similar to that found in cytochrome c(163). Moreover, studies with mutant hemoglobins in which lysines situated near the opening to the heme crevice are converted to glutamate residues have established the importance of these lysines in the reaction with b_5(167). With this information in hand and the aid of a computer graphics system, the b_5-methemoglobin complex shown in Fig. 13 was proposed(163). This model is remarkably similar to the b_5-c model in that the same b_5 carboxylates are involved, but an even more striking similarity is that the hemes are coplanar and only 8 Å apart.

A third short electron transfer distance complex has been proposed for the nonphysiological reaction between flavodoxin and cytochrome c(168). Charge interactions between flavodoxin and cytochrome c are once again very similar to those found in the other two models, but in this case there is direct contact between redox

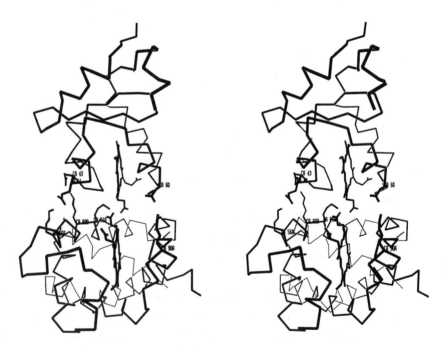

FIG. 13. Stereoscopic view of the hypothetical hemoglobin α chain (bottom) - cytochrome b_5 (top) complex proposed by Poulos and Mauk(163). Residues involved in intermolecular ionic contacts are labeled.

groups(168). The flavodoxin isoalloxazine ring is only 3.4 Å from the cytochrome c heme enabling electron transfer to occur via a simple outer sphere mechanism(169).

d. CCP-Cytochrome c Complex

The fourth, and last, hypothetical model to be considered is the CCP-c complex which represents the only example of a long distance electron transfer complex. The initial hypothetical CCP-c complex was based on the unrefined 2.5 Å CCP(151, 170) and 2.0 Å tuna cytochrome c structures(171). Since then, both structures have been refined, CCP to 1.7 Å, R = .21 (to be published) and cytochrome c to 1.8 Å, R = .20(172). As a result, we have reexamined the CCP-c complex on the computer graphics system using refined structures and the resulting hypothetical model is shown in Fig. 14. Figure 15 shows, in more detail, the points of hydrogen bonding and/or ionic contacts.

The overall complementary charge distribution and interaction domains remain the same as in our earlier study(151). In determining potential charge-charge interactions, we assumed, as before, that the sidechains of charged residues are free to rotate in order to optimize hydrogen bonding distances and geometry as long as a fully extended conformation is maintained and no unfavorable inter- or intramolecular contacts result. As shown in Table 2, several of the intermolecular contacts observed in the original hypothetical model are still found in the refined structures though we now observe additional contacts (see Table 2). In addition, the Asp 79 CCP-Lys 27 cytochrome c interaction found previously is absent in the present model since Asp 79 in CCP was incorrectly positioned in the MIR model. One other important change is the interaction between His 181 in CCP and Phe 82 in cytochrome c (Fig. 14). While the parallel arrangement of the His 181 and Phe 82 sidechains was correctly predicted in our previous study(151), the exact positioning of the His 181 sidechain was not clear in the MIR maps. At the time, we knew that the outermost heme propionate (pyrrole IV) interacts strongly with an amino acid sidechain, but we were not sure if the key residue was Thr 180 or His 181. It is now clear that His 181 is the residue in question. It is positioned between both hemes (Fig. 14) about 7 Å from, and approximately parallel to, the invariant phenyl ring of Phe 82 in cytochrome c.

It seems likely that ion pairing or hydrogen bonding contributes very little to the overall negative free energy of complex formation. The driving force in form-

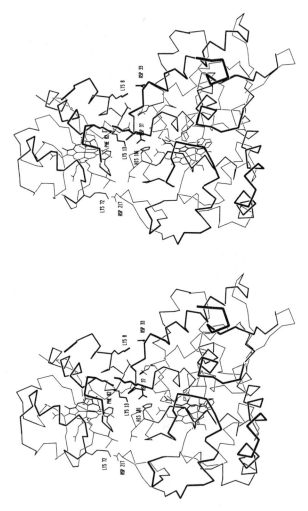

FIG. 14. Stereoscopic representation of the hypothetical CCP (bottom) - cytochrome *c* (top) complex. For clarity only a few key residues involved in intermolecular contacts are labeled.

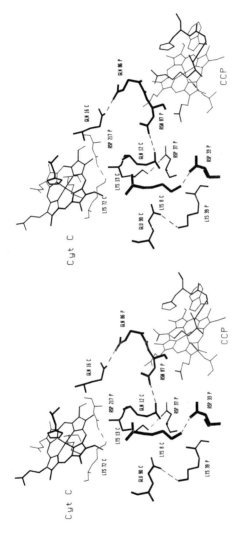

FIG. 15. Stereoscopic model of the CCP-cytochrome *c* complex showing the points of ionic and/or hydrogen bonding contacts. CCP residues are labeled with a P and cytochrome *c* with a C.

TABLE 2

Hydrogen-bonding and/or Ionic Interactions in the Cytochrome c Peroxidase · Tuna Cytochrome c Complex. Distances Were Measured Between Atomic Centers.

Cytochrome c peroxidase	Tuna cytochrome c	Corresponding residue in yeast iso-1-cytochrome	Interaction present in previous study	Distance	Comments
Asp 33	Lys 8	Lys 10	NO	2.8 Å	
Asp 34	Lys 87	Lys 92	YES	4.7 Å.	Too far for strong interaction
Asp 37	Lys 13	Arg 18	YES	2.8 Å	
Gln 86	Gln 16	Gln 21	YES	2.9 Å	
Asn 87	Gln 12	Thr 17	YES	2.8 Å	
Asp 217	Lys 72	Lys 77	YES	2.8 Å	
His 181	Phe 82	Phe 87	YES	7 Å	Imidazole and phenyl rings are nearly parallel
Leu 182	Ile 81	Ala 86	NO	3.0 Å	
heme edge	heme edge	--	--	17.8 Å	
Fe	Fe	--	--	25.1 Å	

ing such intermolecular complexes is probably the entropically favored release of solvent molecules at the intermolecular (heme-heme) interface and other nonpolar contact regions(160). One such nonpolar contact in the CCP-*c* complex is the Leu 182 CCP-Ile 81 cytochrome *c* contact (Table 2). Interestingly, an exposed hydrophobic residue at position 81 is found in all cytochromes *c*(174) suggesting that an exposed hydrophobic area at this position is important in forming intermolecular electron transfer complexes(175). The role of ion pairing may be, therefore, to correctly orient the redox pairs and bring the hemes into parallel alignment but not necessarily to provide the energetic driving force in forming the complex.

Overall, the CCP-*c* complex is very similar to the short distance complexes described earlier in its degree of surface complementarity. The most noteworthy similarity, however, is that optimization of surface interactions results in the parallel alignment of the hemes. This hardly seems coincidental and may well be an important, if not critical, feature in the electron transfer process. On the other hand, the most notable difference between the CCP-*c* complex and the other models is that the heme edges are ~17-18 Å apart, approximately twice the distance found in the b_5-*c* and b_5-hemoglobin models. We emphasize that this is the closest possible approach of the two hemes owing to the location of the CCP heme well within the protein.

Experimental Support of the Hypothetical CCP-*c* Model

The available experimental data supports the hypothetical CCP-*c* model. The 17-18 Å distance between hemes in the model is consistent with both fluorescent energy transfer(135) and NMR(176) estimates. A more definitive test, however, would be to map the reactive surface of CCP involved in recognition of cytochrome *c* using chemical modification methods in a manner similar to that employed in mapping the recognition domain of cytochrome *c*. Though still not complete, the data from such an approach is beginning to emerge. Poulos *et al.*(177) and Waldemeyer *et al.*(145) found that CCP carboxylates are indeed required for reaction with cytochrome *c*. It was also found that reaction of CCP with a carboxylate specific water soluble carbodiimide in the presence of cytochrome *c* results in a 1:1 CCP-*c* covalently crosslinked product in at least 30% yield(145). Presumably the crosslinks arise from the reaction of cytochrome *c* lysine amino groups with carbodiimide activated CCP carboxylates. Isolation of cyanogen bromide fragments derived from the CCP-*c* covalent complex demonstrated that those regions of the molecules predicted to interact on the basis of the hypothetical model do indeed crosslink. Nevertheless, the peptides isolated were rather large and further sequencing will be required to identify the precise sites of interaction.

Using a somewhat different approach, Bisson and Capaldi(178) prepared a monosubstituted cytochrome *c* derivative in which a photoactive arylazido group was covalently attached to Lys 13. After crosslinking a 1:1 mixture of CCP and the Lys 13 derivative, a crosslinked peptide was isolated and it was found that Lys 13 of cyto-chrome *c* crosslinked with the 32-48 region in CCP. This is precisely where the model predicts a crosslink should occur since in the hypothetical model Lys 13 interacts with Asp 37.

The Asp 37 region is particularly interesting. Residues 32-40 in CCP are situated on the surface and connect two long α helical segments with residues 36-40 forming the unusual left handed helical turn described earlier. The sequence of the 32-40 region is Glu-Asp-Asp-Glu-Tyr-Asp-Asn-Tyr-Ile so that five of the nine residues in this stretch are Glu or Asp. Therefore, this loop presents a cluster of negative charges for interaction with the Lys 13 cluster of positive charges in cytochrome *c*. In fact, Bechtold *et al.*(179) found that in the presence of cytochrome *c*, the carboxylates in the 32-48 tryptic peptide of CCP are 5 times less reactive toward carbodiimides than in the absence of cytochrome *c*, further demonstrating that cytochrome *c* interacts with this region of the CCP molecule.

To briefly summarize, the CCP-*c* hypothetical model is fully compatible with the available experimental data, strongly indicating that the model is a very close approximation to the functionally relevant electron transfer complex.

CCP-*c* Electron Transfer Circuit

In the short distance electron transfer complexes we have considered there is no intervening protein between the two redox groups. In our long distance example, the CCP-*c* complex, the situation is quite different. An intriguing system of conjugated and hydrogen bonding interactions involving amino acid sidechains is situated between the two hemes near the center of the interaction domain (see Fig. 16 for a closeup view). His 181 of CCP is situated at the molecular surface near the center of the inter-molecular interface. It serves to complete an intramolecular hydrogen bonding network involving the outermost heme propionate and Arg 48 which connects the surface of CCP with the distal side of the heme. As we described earlier, His 181 also hydrogen bonds with Asp 37 and we know from chemical modification studies that Asp 37 is situated in the surface loop that crosslinks with cytochrome c(178). Irrespective of whether or not the proposed role for His 181 in the electron transfer reaction to be discussed below is correct, His 181 must be situated near the intermolecular interface.

In the model, His 181 is the closest conjugated CCP group to the cytochrome c heme. His 181 is ~7 Å from the invariant Phe 82 of cytochrome c while Phe 82 is just below and parallel to the cytochrome c heme (see Fig. 16). Note, too, that all conjugated and aromatic groups between and including the two hemes are approximately parallel to one another. Such an arrangement of parallel conjugated groups, as well as the intramolecular hydrogen bonding network connecting His 181 with the distal side of the CCP heme crevice, may well serve as an effective conduit for the transfer of an electron and a proton from cytochrome c to CCP Compound I. The longest distance the electron need traverse directly through the bridge is about 7 Å which is much closer to the distance involved in the cytochrome c-b_5 and b_5-hemoglobin models.

Also shown in Fig. 16 is the Lys 13-Asp 37 ion pair which is located near the center of the intermolecular interface and the proposed electron transfer bridge. It is interesting to consider the possibility that this pair helps trigger the electron transfer reaction by altering the electronic environment of the heme-heme conduit. Precisely how this might occur is open to speculation but one possibility is that ion pairing or hydrogen bonding between Lys 13 and Asp 37 weakens the intramolecular Asp 37-His 181 hydrogen bond in CCP thereby enabling His 181 to serve as a more effective electron acceptor.

The involvement of both His 181 and Arg 48 in the electron transfer circuit is particularly interesting since histidine and arginine are unique among the amino acids in sharing three properties which are especially well suited for participation in an electron-proton circuit: 1) both are conjugated; 2) both are capable of hydrogen bonding; and 3) both the imidazole and guanidinium groups exhibit internal two-fold symmetry so an electron and/or proton entering one side of the imidazole or guanidinium group can exit from the opposite side without the need for significant sidechain movements.

While to the best of our knowledge arginine has not been previously implicated in an electron transfer reaction, histidine has. Pulse radiolysis studies on the reduction of copper in ceruloplasmin have implied the involvement of histidine(180).

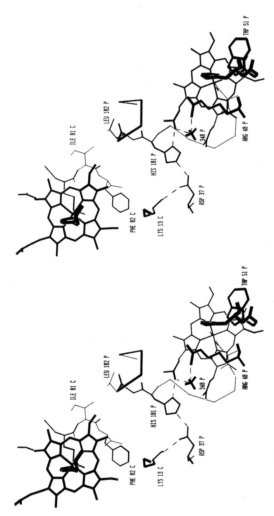

FIG. 16 Stereoscopic model of the CCP-cytochrome c electron transfer conduit looking down on the distal surface of the CCP heme. CCP residues are labeled with a P and cytochrome c residues with a C. Note the intramolecular hydrogen bonding network in CCP involving Arg 48, water 348, pyrrole IV propionate, His 181, and Asp 37, and the intermolecular ionic interaction between Lys 13C and Asp 37P.

An example more closely related to the CCP-c system is the reduction of the bacterial copper protein, azurin. Both NMR(181) and crystallographic data(182) indicate that a histidine residue in *Alcaligenes faecalis* azurin serves as a conjugated path for electron transfer to the copper atom(181).

The requirement of conjugated and hydrogen bonding groups serving as a path for electron transfer is consistent with the more thoroughly understood inorganic and organic models. Electron transfer through conjugated groups is well known(183-188) as is the participation of hydrogen bonds in outer sphere electron transfer processes(189). In fact, hydrogen atom transfer is thought to occur over very large distances ($> 100\,\text{Å}$) via a system of hydrogen bonded water molecules(190).

Whether or not the proposed electron transfer circuit is critical in the CCP-c reaction has not been rigorously tested but the requirement of the heme propionates, which form an important link in the proposed electron transfer circuit, has been established. Asakura and Yonetani(191) and Mochan(192) studied the properties of CCP in which the heme was replaced with hemes having esterified propionates. The esterified enzyme 1) reacts normally with H_2O_2 to give Compound I, 2) utilizes artifical electron donors like ferrocyanide as a substrate at about 50% the rate of the native enzyme, 3) forms a normal 1:1 complex with cytochrome c, but 4) the electron transfer reaction from cytochrome c is more than 99% inhibited. Therefore, even though cytochrome c can bind to the modified CCP, the electron transfer reaction is blocked.

That esterification of the heme propionates blocks electron transfer and should, therefore, disrupt the heme-heme conduit supports the view that the proposed electron transfer conduit is important for the efficient transfer of electrons. We suggest that not only are complementary surfaces required to bring the donor and acceptor hemes into parallel alignment but a specific electron transfer conduit is also required when the distance between redox centers is large.

e. Electron Transfer Mechanisms

While the hypothetical models we have been considering offer stereochemical constraints on possible electron transfer mechanisms, they do not in themselves address the question of precisely how the transfer of electrons proceeds. We conclude by briefly considering some possibilities.

When two redox partners form an electron transfer complex, formation of a conduction band could result enabling the electron to freely pass between donor and acceptor redox centers. For such a mechanism to be useful in biological systems, the energy gap between the valence and conduction bands must be sufficiently narrow to allow the electron to be elevated to the conduction band under normal physiological conditions. However, experimental estimates for the activation energies involved in this process are about 2 to 3 eV (46 to 70 kcal) in proteins(193-196) and theoretical estimates for the energy gap in polyglycine are on the order of 16.5 eV(197) which is far too high to be of much use in biology. Even so, one could argue that the precise ordering of conjugated groups between donor and acceptor redox centers might lower the activation energy enough to enable a conduction band mechanism to operate. This too, however, is an unlikely possibility. Photoemission spectroscopy of various organic polymers shows that the spectra of the polymer having closely spaced aromatic groups (i.e. styrene) is very similar to that of the monomer (benzene) and there is no evidence for a conduction band(198). In other words, " $\cdots$ electrons or holes injected into a polymer form molecular radicals (local anions and cations, respectively) rather than extended mobile states like those found in covalent semiconductors"(198). Therefore, we can expect electron transfer to occur between localized sites in biological systems. A localized site electron transfer process is actually prefered since it insures unidirectional and specific transfer in biological systems where the reaction is downhill with respect to free energy(199). This is not the case in a conduction band process where the electron is freely mobile(199).

In order for the electron to transfer between localized sites it must either attain enough energy to hop over the activation energy barrier or burrow through the barrier via quantum mechanical tunneling. Tunneling mechanisms are often preferred in biological systems, especially for long distance reactions, since the expected activation energies are high.

In general we can express the rate of electron transfer, R, via a tunneling mechanism as follows(200)

$$R = fT$$

where f = the frequency with which the electron encounters the barrier and T = the tunneling matrix. T is a measure of the tunneling probability and is given by

$$T = Ke^{[-\alpha[8m(V-E)]^{1/2}/\hbar]}$$ (1)

where α = barrier width, m = mass of an electron, V = barrier height, E = energy of the electron, $\hbar$ = Planck's constant / 2π and $K = 16 / [2 + (V-E)/E + E/(V-E)(200)$. Estimates of the various unknown parameters have been made to determine the distance between redox centers. In the well known examples of electron transfer from cytochrome c to the photoxidized reaction center in *Chromatium* (200) and *Rhodopseudomonas spheroides* (201) distance estimates are quite long (> 20 Å). However, Hopfield(202) has raised some serious objections to these earlier treatments, especially criticizing the assumption made that the pre-exponential frequency factor in Eq. (1) is a constant and reaction independent.

Hopfield offers an interesting alternative by constructing an analogy between the electron transfer process and excitation transfer spectroscopy(202). The distribution of energies of an electron in either the donor or acceptor define a gaussian band similar to an absorption spectral band. These bands interact only weakly when the distance between redox centers is large, so the probability of tunneling will be low. However, just as spectral band widths increase owing to thermally activated nuclear vibration, the electron transfer bands also broaden with thermal nuclear motions thereby increasing band overlap and the probability of an electron tunneling from donor to acceptor. The required nuclear vibrational energy is obtained from the conservation of kinetic energy when the donor and acceptor molecules collide to form a complex(173). The most important result emerging from Hopfield's treatment is that the estimate for the electron transfer distance in the bacterial photosynthetic system is much shorter (~ 8 Å) than earlier estimates (> 20 Å).

The shorter distance predicted from Hopfield's treatment is more consistent with the short distance electron transfer complexes described earlier, and site to site transfer through the electron transfer circuit in the CCP-c system. Additionally, Hopfield's treatment has been used to give structurally reasonable electron transfer distances in several metalloprotein redox reactions where one of the reactants is a small molecule complex of iron, cobalt, or ruthenium(203).

Certainly such theoretical treatments are useful in conceptualizing how biological electron transfer reactions proceed, but in general they suffer from the lack of quantitative information regarding the many unknown parameters which must be estimated, especially those regarding distance, orientation, and the medium through which the electron must be transfered. What is required are simple experimental systems which parallel or closely mimic physiological systems and where the various structural parameters are known. Toward this end, an interesting model system has been developed recently in which pentaaminoruthenium is covalently linked to histidines in cytochrome c(157) or myoglobin(204) enabling accurate determination of intermolecular electron transfer rates between the ruthenium and heme iron atoms. Since both the protein structures and the site(s) of ruthenium attachment are known, accurate correlations between electron transfer rates and distances and/or the nature of the protein matrix separating the metal atoms are possible. Additionally, the hypothetical models described previously offer considerable insight into the geometry of interaction between donor and acceptor redox centers in addition to the electron transfer medium. Hopefully, such an experimental approach will introduce realistic structural constraints into electron transfer theory resulting in theoretical treatments that come closer to satisfying empirical observations.

Whether or not tunneling is the preferred mechanism, there are few problems in visualizing that coplanar alignment of hemes in the short electron transfer complexes we considered in Section (V.c) allows direct transfer between heme edges. However, the CCP-c system is more complex. Especially troublesome if site to site transfer occurs through the proposed conduit, is that formation of protein centered radicals is required, at least transiently, and that the energy barrier for the production of such radicals is prohibitively large(205). However, as Winfield argued several years ago(183) and as we have emphasized earlier in this article, CCP, and probably other heme enzymes as well, have heme and/or amino acid centered radicals as a result of reaction with peroxides or O_2. Once a "hole" has been introduced in this way, transfer of electrons through protein groups need not present a significant energy barrier, leaving open the possibility for mechanisms which involve hopping over the barrier as well as tunneling through it.

Two other mechanisms for facilitating the electron transfer process are available in biological systems. First, the metal atom and its ligands may be held by the protein in a configuration resembling the activated complex; that is, held in a configuration midway between the oxidized and reduced states(206), often called the entatic state(207). Second, the free energy of complex formation is utilized to adjust the nuclear coordinates of the electron transfer groups in a direction closer to the activated complex. This latter mechanism requires that intermolecular surface interactions between donor and acceptor proteins results in readjustment of energy levels at the active center and/or electron transfer groups. In site to site transfer through the protein, readjustment of electron transfer groups could occur between electron transfer events, giving rise to as many activated complexes as there are localized sites.

Both mechanisms are feasible. In model systems, electron transfer between cupric (Cu^{2+}) and cuprous (Cu^{1+}) complexes is faster when neither the donor nor acceptor metal atoms are in a tetrahedral configuration (preferred by Cu^{1+}) or a planar configuration (preferred by Cu^{2+}) but are held midway between the two extremes(208). Furthermore, evidence exists that in horseradish peroxidase (209) and cytochrome c'(210) interactions between the iron atom and protein adjust the spin state of the iron to be neither low nor high, but a quantum admixture of both spin states(209). As Maltempo has argued for peroxidase(209), holding the iron in such a spin state will facilitate its oxidation to low spin Fe(IV) in the formation of Compound I because more energy is required to go from high spin Fe(III) to low spin Fe(IV) than from high/low spin quantum admixture Fe(III) to low spin Fe(IV)(209).

One of the best examples for communication between surface interactions and the interior redox center is the $P450_{CAM}$-putidaredoxin reaction. Putidaredoxin is an effector molecule as well as a reductant of $P450_{CAM}(36)$. That is, efficient formation of the hydroxycamphor requires the interaction of putidaredoxin with $P450_{CAM}(36)$. In fact, recent kinetic data indicates that formation of the 1:1 putidaredoxin-$P450_{CAM}$ followed by a conformational change prior to electron transfer(211). Therefore, it appears that a switch operates to communicate intermolecular surface interactions between putidaredoxin and $P450_{CAM}$ to the site of camphor hydroxylation which we expect, on the basis of the X-ray data, lies on the interior side of the heme, well removed from the molecular surface.

The hypothetical CCP-c complex also suggests possible structural switches which may effect the electron transfer process. One such switch involves Lys 13 on cytochrome c. If indeed Lys 13 forms an intermolecular ion pair with CCP as we believe it does, then the intramolecular Lys 13-Glu 90 ion pair in cytochrome c is either weakened or disrupted. Disruption of the Lys 13-Glu 90 pair could have a significant effect on the electron transfer reaction since Osheroff et al. (212) have shown that the

Lys 13-Glu 90 ion pair is important for stabilization of the cytochrome c heme crevice and effects the electronic properties of the heme. In addition, we have obtained crystallographic evidence for a ligand induced conformational change in CCP involving the region of the molecule we believe interacts with cytochrome c. A 2.5 Å difference Fourier of CCP complexed with nitric oxide (NO), a low spin ligand(213), shows a clear and obvious conformational change involving residues 179-194(214, 215). This means that changes in the electronic structure of the heme result in a structural rearrangement at the surface of the protein well removed from the site of NO binding. Precisely how the structural change is communicated from the distal side of the heme to the molecular surface is not clear. Nevertheless, the residues involved in the ligand induced conformational change form part of the antiparallel β pair on the surface of CCP situated at the center of the proposed CCP-cytochrome c interaction domain. Furthermore, His 181, which forms an important link in the heme-heme conduit, is also involved in the NO induced conformational change. That this particular region is sensitive to electronic and structural changes in the heme crevice is intriguing if the reverse is also true; that is, formation of the CCP-c complex induces changes at the intermolecular interface resulting in alterations of the respective heme crevices and electron transfer groups which then allows the electron transfer reaction to proceed.

Acknowledgements: Crystallographic work on CCP and P450 was supported by NSF Grant PCM79-14595 and computing and graphics by NIH Grant RR 00757. We would like to thank Muriel Finzel for assistance in the preparation of the manuscript and Dr. Jane Burridge for a critical review.

REFERENCES

1.   Welinder, K. G. and Mazza, G. *Eur. J. Biochem.* **73**, 353-358 (1977)

2.   Mazza, G. and Welinder, K. G. *Eur. J. Biochem.* **108**, 481-489 (1980)

3.   Welinder, K. G. *FEBS Letters* **72**, 19-23 (1976)

4.   Takio, K., Titani, K., Ericsson, L. H., and Yonetani, T. *Arch. Biochem. Biophys.* **203**, 615-619 (1980)

5.   Poulos, T. L. and Kraut, J. *J. Biol. Chem.* **255**, 8199-8205 (1980)

6.   Jones, P. and Dunford, H. B. *J. Theor. Biol.* **69**, 457-470 (1977)

7. Yamazaki, I., Sano, H., Nakajima, R., and Yokota, K-n. *Biochem. Biophys. Res. Comm.* **31**, 932-937 (1968)

8. Morrison, M. and Schonbaum, G. R. *Ann. Rev. Biochem.* **45**, 861-888 (1976)

9. Hewson, W. D. and Hager, L. P. in *The Porphyrins,* (D. Dolphin, ed.) vol.7, pt.B pp. 295-332, Academic Press, 1979

10. Henson, P. M., Giusberg, M. H., and Morrison, D. C. *Cell. Surj. Rev.* **5**, 407-508 (1978)

11. Root, R. K., Metcalf, J., Oshino, N., and Chance, B. *J. Clin. Invest.* **55**, 945-955 (1975)

12. Levine, P. H., Weinger, R. S., Simon, J., Scoon, K. L., and Krinsky, N. I. *J. Clin. Invest.* **57**, 955-963 (1976)

13. Klebanoff, S. J. and Clark, R. H. in *The Neutrophil: Function and Clinical Disorders,* Elsevier, North-Holland, Amsterdam, 1978

14. Babior, B. M. *N. Eng. J. Med.* **298**, 659-666; 721-725 (1978)

15. Thomas, E. L. *Infect. Immun.* **23**, 522-531 (1979)

16. Zgliczynski, J. M. and Stelmaszynska, T. *Eur. J. Biochem.* **56**, 157-162 (1975)

17. Albrich, J. M., McCarthy, C. A., and Hurst, J. K. *Proc. Natl. Acad. Sci. U.S.A.* **78**, 210-214 (1981)

18. Taurog, A. *Hanb. Physiol.* **3**, 101-134 (1974)

19. Taurog, A., Lothrop, M. L., and Estabrook, R. W. *Arch. Biochem. Biophys.* **139**, 221-229 (1970)

20. Bayse, G. S., Michaels, A. W., and Morrison, M. *Biochim. Biophys. Acta.* **284**, 34-42 (1972)

21. Kuehl, F. A. Jr., Humes, J. L., Ham, E. A., Egan, R. W., and Dougherty, H. W. *Advan. In Prostaglandin and Thromboxane Res.* **6**, 77-86 (1980)

22. Ts'o, H. V. Gilbain, P. O. eds. in *Polycyclic Hydrocarbons and Cancer,* Academic Press, New York, 1978

23. Hayashi, O. ed. in *Molec. Mech. of Oxygen Act.,* Academic Press, New York, 1974

24. Gunsalus, I. C., Pederson, T. C., and Sligar, S. G. *Ann. Rev. Biochem.* **44**, 377-407 (1975)

25. Nebert, D. W., Eisen, H. J., Negishi, M., Lang, M. A., Hjelmeland, L. M., and Okey, A. B. *Ann. Rev. Pharmacol. Toxicol.* **21**, 431-462 (1981)

26. Coon, M. J. and White, R. E. in *Dioxygen Binding and Activation by Metal Centers*,
     (T. G. Spiro, ed.) pp. 73-123, John Wiley and Sons, New York, 1980

27. Ullrich, V. *Angew. Chimie, intern edit.* **11**, 701-712 (1972)

28. Tomeszewski, J. E., Jerina, D. M., and Daly, J. W. *Ann. Reports Medicin. Chem.*
     9, 290-299 (1974)

29. Ullrich, V. *Topics in Curr. Chem.* **83**, 67-104 (1979)

30. Omura, T., Sanders, E., Estabrook, R. W., Cooper, D. Y., and Rosenthal, O.
     Arch. Biochem. Biophys. **117**, 660-666 (1966)

31. Kulkoski, J. A. and Ghazarian, J. G. *Biochem. J.* **177**, 673-678 (1979)

32. Atsuta, Y. and Okuda, K. *J. Biol. Chem.* **253**, 4653-4658 (1978)

33. Mitani, F. *Mol. Cell. Endocrinol.* **13**, 213-227 (1979)

34. Kimura, T., Parcells, J. H., and Wang, H. P. *Methods in Enzymol.* **52**, 132-142
     (1978)

35. Estabrook, R. W., Cooper, D. Y., and Rosenthal, O. *Biochem.* **7**, 338, 741-755
     (1963)

36. Gunsalus, I. C., Meeks, J. R., Lipscomb, J. D., Debrunner, P. G., and Munck, E.
     in *Molec. Mech. of Oxygen Act.*, (O. Hayoshi,ed.) pp. 559-613, Academic Press,
     New York, 1974

37. Holm, R. H., Tang, S. C., Koch, S., Papaefthymiou, G. C., Foner, S., Frankel, R.
     R., and Ibers, J. A. *Adv. Exp. Med. Biol.* **74**, 321-334 (1976)

38. Ruf, H. H., Wende, P., and Ullrich, V. *J. Inorg. Chem.* **11**, 189-204 (1979)

39. Stern, J. O. and Peisach, J. *J. Biol. Chem.* **249**, 7495-7498 (1974)

40. Wagner, G. C., Gunsalus, I. C., Wang, M.-Y., and Hoffman, B. *J. Biol. Chem.*
     256, 6266-6273 (1981)

41. Champion, P. M., Stellard, B. R., Wagner, G. C., and Gunsalus, I. C. *J. Amer.*
     Chem. Soc. **104**, 5469-5472 (1982)

42. Dolphin, D., James, B. R., and Welborn, H. C. *Biochem. Biophys. Res. Comm.* **88**,
     415-421 (1979)

43. Dawson, J. H., Holm, R. H., Trudell, J. R., Barth, G., Linder, R. E., Bunnen-
     berg, E., Djerassi, C., and Tang, S. C. *J. Amer. Chem. Soc.* **98**, 3707-3709 (1976)

44. Collman, J. P. and Sorrell, T. N. in *ACS Symposium Series No.44, Drug Metabol-*
     ism Concepts, (D. M. Jerina, ed.) pp. 27-45, Amer. Chem. Soc., Washington, D.
     C., 1977

45. Cramer, S. P., Dawson, J. H., Hodgson, K. O., and Hager, L. P. *J. Amer. Chem. Soc.* **100**, 7282-7290 (1978)

46. Chang, C. K. and Dolphin, D. *J. Amer. Chem. Soc.* **97**, 5948-5950 (1975)

47. Dawson, J. H., Andersson, L. A., Davis, I. M., and Hahn, J. E. in *Biochemistry, Biophysics, and Regulation of Cyt. P450*, (J. A. Gustafsson, J. Carlstedt-Duke, A. Mode, and J. Rafter,eds.) pp. 565-572, Elsevier, Amsterdam, 1980

48. Collman, J. P., Sorrell, T. N., Dawson, J. H., Trudell, J. R., Bunnenberg, E., and Djerassi, C. *Proc. Natl. Acad. Sci. U.S.A.* **73**, 6-10 (1976)

49. Chang, C. K. and Dolphin, D. *Proc. Natl. Acad. Sci. U.S.A.* **73**, 3338-3342 (1976)

50. Hahn, J. E., Hodgson, K. O., Andersson, L. A., and Dawson, J. H. *J. Biol. Chem.* **257**, 10934-10941 (1982)

51. Fujii-Kuriyama, Y., Mizukami, Y., Kawajiiri, K., Sogawa, K., and Muramatsu, M. *Proc. Natl. Acad. Sci. U.S.A.* **79**, 2793-2797 (1982)

52. Haniu, M., Armes, L. G., Tanaka, M., Yasunobu, K. T., Shastry, B. S., Wagner, G. C., and Gunsalus, I. C. *Biochem. Biophys. Res. Comm.* **105**, 889-894 (1982)

53. Heinemann, F. S. and Ozols, J. *J. Biol. Chem.* **257**, 1498-1499 (1982)

54. Black, S. D., Tarr, G. E., and Coon, M. J. *J. Biol. Chem.* **257**, 14616-14619 (1982)

55. Ozols, J., Henemann, F. S., and Johnson, E. F. *J. Biol. Chem.* **256**, 11405-11408 (1981)

56. Ishimura, Y., Ullrich, V., and Peterson, J. A. *Biochem. Biophys. Res. Comm.* **42**, 140-146 (1971)

57. Sharrock, M., Debrunner, P., Schulz, C., Lipscomb, J., Marshall, V., and Gunsalus, I. C. *Biochim. Biophys. Acta.* **420**, 8-26 (1976)

58. Lipscomb, J., Sligar, S., Namtvedt, M., and Gunsalus, I. C. *J. Biol. Chem.* **251**, 1116-1124 (1976)

59. Sligar, S. G., Shastry, B. S., and Gunsalus, I. C. in *Microsomes and Drug Oxidation*, (V. Ullrich, I. Roots, A. Hildebrandt, R. W. Estabrook, and A. H. Cooney,eds.) pp. 202-209, Pergamon Press, Oxford, 1977

60. Kadlubar, F. F., Morton, K. C., and Ziegler, D. M. *Biochem. Biophys. Res. Comm.* **54**, 1255-1261 (1973)

61. White, R. E., Sligar, S. G., and Coon, M. J. *J. Biol. Chem.* **255**, 11108-11111 (1980)

62. Blake, R. C. and Coon, M. J. *J. Biol. Chem.* **256**, 5755-5763 (1981)

63. Blake, R. C. and Coon, M. J. *J. Biol. Chem.* **256**, 12127-12133 (1981)

64. Hamilton, G. A. in *Molecular Mechanisms of Oxygen Activation*, (O. Hayashi,ed.) pp. 405-451, Academic Press, 1974

65. Hamilton, G. A. *Advan. in Enzymol.* **32**, 55-96 (1969)

66. Hamilton, G. A., Hanitin, J. W., and Friedman, J. P. *J. Amer. Chem. Soc.* **88**, 5269-5272 (1966)

67. Hamilton, G. A., Friedman, J. P., and Campbell, P. M. *J. Amer. Chem. Soc.* **88**, 5266-5268 (1966)

68. Warshel, A. and Levitt, M. *J. Mol. Biol.* **103**, 227-249 (1976)

69. Warshel, A. *Chem. Phys. Lett.* **53**, 454-458 (1978)

70. Vernon, C. A. *Proc. R. Soc. Lon., Ser. B* **167**, 389-401 (1967)

71. Perutz, M. F. *Proc. R. Soc. Lon., Ser. B* **167**, 448 (1967)

72. Warshel, A. *Proc. Natl. Acad. Sci. U.S.A.* **75**, 5250-5254 (1978)

73. Walsh, C. in *Enzyme Reaction Mechanisms*, W. H. Freeman and Co., San Francisco, 1979

74. Azzi, A. *Biochim. Biophys. Acta.* **594**, 231-252 (1980)

75. Henderson, R., Capaldi, R. A., and Leigh, J. S. *J. Mol. Biol.* **112**, 631-648 (1977)

76. Mason, T. L. and Schatz, G. *J. Biol. Chem.* **248**, 1355-1360 (1973)

77. Schatz, G. and Mason, T. L. *Annu. Rev. Biochem.* **43**, 51-87 (1974)

78. Lewin, A., Gregor, I., Mason, T. L., Nelson, N., and Schatz, G. *Proc. Natl. Acad. Sci. U.S.A.* **77**, 3998-4002 (1980)

79. Mihara, K. and Blobel, G. *Proc. Natl. Acad. Sci. U.S.A.* **77**, 4160-4164 (1980)

80. Steffens, G. J. and Buse, G. *Hoppe-Seyler's Z. Physiol. Chem.* **360**, 613-619 (1979)

81. Winter, D. B., Bruynincky, W. J., Foulke, F. G., Grinich, N. P., and Mason, H. S. *J. Biol. Chem.* **255**, 11408-11414 (1980)

82. Carbral, F., Solioz, M., Deters, D., Rudin, Y., Schatz, G., Clavilier, L., Groudinsky, O., and Slonimski, P. in *Genetics and Biogenesis of Mitochondria*, (W. Bandlow, R. J. Schweyen, K. Wolf, and F. Kaudewitz,eds.) pp. 401-413, deGruyter, Berlin, 1977

83. Ludwig, B. *Biochim. Biophys. Acta.* **594**, 177-189 (1980)

84. Chance, B., Saronio, C., Leigh, J. S., and Waring, A. in *Tunneling in Biological Systems*, (B. Chance, D. C. DeVault, H. Fraunfelder, R. A. Marcus, J. R. Schrieffer, and N. Sutin,eds.) pp. 483-511, Academic Press, 1979

85. Bickar, D., Bonaventura, J., and Bonaventura, C. *Biochem.* **21**, 2661-2666 (1982)

86. Orii, Y. *J. Biol. Chem.* **257**, 9246-9248 (1982)

87. Kaput, J., Glotz, S., and Blobel, G. *J. Biol. Chem.* **257**, 15054-15058 (1982)

88. Ten Eyck, L. F. in *The Porphyrns*, (D. Dolphin, ed.) VII Pt.B Chap.10, Academic Press, 1979

89. Matthews, F. S., Czerwinski, E. W., and Argos, P. in *The Porphyrins*, (D. Dolphin, ed.) VII Pt.B Chap.3, Academic Press, 1979

90. Matthews, F. S., Bethge, P. H., and Czerwinski, E. W. *J. Biol. Chem.* **254**, 1699-1706 (1979)

91. Weber, P. C., Bartch, R. G., Cusanovich, M. A., Hamlin, R. C., Howard, A., Jordan, S. R., Kamen, M. D., Meyer, T. E., Weatherford, D. W., Xuong, Ng.-h, and Salemme, F. R. *Nature* **286**, 302-304 (1980)

92. Miles, E. W. *Methods in Enzym.* **47 pt.E**, 431-432 (1977)

93. Poulos, T. L. *unpublished results*

94. Izuka, T., Katani, M., and Yonetani, T. *J. Biol. Chem.* **246**, 4731-4736 (1971)

95. Saterlee, J. D., Erman, J. E., LaMar, G. N., Smith, K. M., and Langry, K. C. *Biochim. Biophys. Acta.* **743**, 246-255 (1983)

96. Valentine, J. S., Sheridan, R. P., Allen, L. C., and Kahn, P. C. *Proc. Natl. Acad. Sci. U.S.A.* **76**, 1009-1013 (1979)

97. LaMar, G. N., Ropp, J. S. De, Chako, V. P., Satterlee, J. D., and Erman, J. E. *Biochim. Biophys. Acta.* **708**, 317-325 (1982)

98. Murthy, M. R. N., Reid, T. J., Sicignano, A., Tanaka, N., and Rossmann, M. G. *J. Mol. Biol.* **152**, 465-499 (1981)

99. Poulos, T. L., Perez, M., and Wagner, G. C. *J. Biol. Chem.* **257**, 10427-10429 (1982)

100. Bernstein, F. C., Koetzle, T. F., Williams, G. J. B., Meyer, E. F. Jr., Brice, M. D., Rodgers, J. R., Kennard, O., Shimanouchi, T., and Tasumi, M. *J. Mol. Biol.* **112**, 535-542 (1977)

101. Takano, T. *J. Mol. Biol.* **110**, 537-568 (1977)

166   POULOS AND FINZEL

102. Lee, W. A., D. Phil. Thesis, Univ. California, San Diego, 1982

103. Conroy, C. W., Tyma, P., Daum, P. H., and Erman, J. E. *Biochim. Biophys. Acta.* **537**, 62-69 (1978)

104. Cassatt, J. C., Marini, C. P., and Bender, J. *Biochem.* **14**, 5470-5475 (1975)

105. Loo, S. and Erman, J. E. *Biochem.* **14**, 3467-3470 (1975)

106. Parsons, S. M. and Raferty, M. A. *Biochem.* **11**, 1623 (1972)

107. Birdsall, B., Gronenborn, A., Hyde, E. I., Clore, G. M., Roberts, G. C. K., Feeny, J., and Burgen, A. S. V. *Biochem.* **21**, 5831-5838 (1982)

108. Ricard, J., Mazza, G., and Williams, R. J. P. *Eur. J. Biochem.* **28**, 566-578 (1972)

109. Hewson, W. D. and Dunford, H. B. *Can. J. Biochem.* **53**, 1928-1932 (1975)

110. Yonetani, T. and Ray, G. S. *J. Biol. Chem.* **241**, 700-706 (1966)

111. Kelley, H. C., Davies, D. M., King, M. J., and Jones, P. *Biochem.* **16**, 3543-3549 (1977)

112. The hydroperoxide atoms are labeled O1 and O2 for easier reference in the text.

113. Jones, P. and Suggett, A. *Biochem.* **110**, 621-629 (1968)

114. Poulos, T. L. in *Molecular Structure and Biological Activity*, (J. E. Griffin, W. L. Duax,eds.) pp. 79-90, Elsevier Biomedical, 1982

115. Job, D. and Jones, P. C. *Eur. J. Biochem.* **86**, 565-572 (1978)

116. Davies, D. M., Jones, P., and Mantle, D. *Biochem. J.* **157**, 247-253 (1976)

117. Jones, P. and Middlemiss, D. N. *Biochem. J.* **130**, 411-415 (1972)

118. Erman, J. E. *Biochem.* **13**, 34-38 (1974)

119. Erman, J. E. *Biochem.* **13**, 39-44 (1974)

120. Dunford, H. B. and Alberty, R. A. *Biochem.* **6**, 447-451 (1967)

121. Ellis, W. D. and Dunford, H. B. *Biochem.* **7**, 2054-2062 (1968)

122. Williams, R. J. P. in *Iron in Biochemistry and Medicine*, (A. Jacobs, M. Worwood,eds.) pp. 183-219, Academic Press, 1974

123. Araiso, T., Miyoshi, K., and Yamazaki, I. *Biochem.* **15**, 3059-3063 (1976)

124. Dolphin, D., Forman, A., Borg, D. C., Fajer, J., and Felton, R. H. *Proc. Natl. Acad. Sci. U.S.A.* **68**, 614-618 (1971)

125. Aasa, R., Vanngard, T., and Dunford, H. B. *Biochim. Biophys. Acta.* **391**, 259-264 (1975)

126. Dunford, H. B. and Stillman, J. S. *Coord. Chem. Rev.* **19**, 187-251 (1976)

127. Yonetani, T., Schelyer, H., and Ehrenberg, A. *J. Biol. Chem.* **241**, 3240-3243 (1966)

128. Erman, J. E. and Yonetani, T. *Biochim. Biophys. Acta.* **393**, 343-349 (1975)

129. Coulson, A. F. W. and Yonetani, T. *Biochem. Biophys. Res. Comm.* **49**, 391-398 (1972)

130. Erman, J. E. and Yonetani, T. *Biochim. Biophys. Acta.* **393**, 350-357 (1975)

131. Hoffman, B. M., Roberts, J. E., Kang, C. H., and Margoliash, E. *J. Biol. Chem.* **256**, 6556-6564 (1981)

132. Fujita, I., Hanson, L. K., Walker, F. A., and Fajer, J. *J. Amer. Chem. Soc.* **105**, 3296-3300 (1983)

133. Vanderkooi, J. M., Landesberg, R., Hayden, G. W., and Owen, G. S. *Eur. J. Biochem.* **81**, 339-347 (1977)

134. Dutton, P. L., Leigh, J. S., Prince, R. C., and Tiede, D. M. in *Tunneling in Biological Systems*, (B. Chance, R. A. Marcus, D. C. DeVault, J. R. Schrieffer, H. Fraunfelder, and N. Sutin,eds.), Academic Press, New York, 1979

135. Leonard, J. J. and Yonetani, T. *Biochem.* **13**, 1465-1468 (1974)

136. Osheroff, N., Brautigan, D. L., and Margoliash, E. *J. Biol. Chem.* **255**, 8245-8251 (1980)

137. Ferguson-Miller, S., Brautigan, D. L., and Margoliash, E. *J. Biol. Chem.* **253**, 149-159 (1978)

138. Rieder, R. and Bosshard, H. R. *J. Biol. Chem.*, 6045-6053 (1978)

139. Smith, H. T., Staudenmayer, N., and Millett, F. *Biochem.* **16**, 4971-4974 (1977)

140. Rieder, R. and Bosshard, H. R. *J. Biol. Chem.* **255**, 4732-4739 (1980)

141. Ahmed, A. J., Smith, H. T., Smith, M. B., and Millett, F. *Biochem.* **17**, 2479-2483 (1978)

142. Speck, H. S., Fergusson-Miller, S., Osheroff, N., and Margoliash, E. *Proc. Natl. Acad. Sci.* **76**, 155-159 (1979)

143. Pettigrew, G. *FEBS Lett.* **86**, 14-16 (1978)

144. Kang, C. H., Brautigan, D. L., Osheroff, N., and Margoliash, E. *J. Biol. Chem.* **253**, 6502-6510 (1978)

145. Waldmeyer, B., Bechtold, R., Bosshard, H. R., and Poulos, T. L. *J. Biol. Chem.* **257**, 6073-6076 (1982)

146. Ng, S., Smith, M. B., Smith, H. T., and Millett, F. *Biochem.* **16**, 4975-4978 (1977)

147. Smith, H. T., Ahmed, A. J., and Millett, F. *J. Biol. Chem.* **256**, 4984-4990 (1981)

148. Speck, S. H., Koppenol, W. H., Dethmers, J. K., Osheroff, N., Margoliash, E., and Rajagopalan, K. V. *J. Biol. Chem.* **256**, 7394-7400 (1981)

149. Salemme, F. R., Kraut, J., and Kamen, M. *J. Biol. Chem.* **248**, 7701-7716 (1973)

150. Salemme, F. R. *J. Mol. Biol.* **102**, 563-568 (1976)

151. Poulos, T. L. and Kraut, J. *J. Biol. Chem.* **255**, 10322-10330 (1980)

152. Ferguson-Miller, S., Brautigan, D. L., and Margoliash, E. in *The Porphyrins*, (D. Dolphin,ed.) vol.VII, pt.B, chap.4, 1979

153. Yonetani, T. *The Enzymes* **13**, 345-361 (1976)

154. Koppenol, W. H. and Margoliash, E. *J. Biol. Chem.* **257**, 4426-4437 (1982)

155. Antalis, T. M. and Palmer, G. *J. Biol. Chem.* **257**, 6194-6206 (1982)

156. Kang, D. S. and Erman, J. E. *J. Biol. Chem.* **257**, 12775-12779 (1982)

157. Yocom, K. M., Winkler, J. R., Nocera, D. G., Bordignon, E., and Gray, H. *Chemica. Scriptu.* **21**, 29-33 (1983)

158. Kraut, J. *Biochem. Soc. Trans.* **9**, 197-203 (1980)

159. Strittmatter, P. in *Rapid Mixing and Sampling Techniques in Biochem.* pp. 71-85, Academic Press, New York, 1964

160. Mauk, M. R., Reid, L. S., and Mauk, A. G. *Biochem.* **21**, 1843-1846 (1982)

161. Guiard, G. and Lederer, F. *Biochimie* **58**, 305-316 (1976)

162. Dailey, H. A. and Strittmatter, P. *J. Biol. Chem.* **254**, 5388-5396 (1979)

163. Poulos, T. L. and Mauk, A. G. *J. Biol. Chem.* (in press)

164. Hultquist, D. E. and Passon, P. G. *Nature* **229**, 252-254 (1971)

165. Sannes, L. G. and Hultquist, D. E. *Biochim. Biophys. Acta.* **544**, 547-554 (1978)

166. Mauk, M. R. and Mauk, A. G. *Biochem.* **21**, 4730-4734 (1982)

167. Gacon, G., Lostanlen, D., Labie, D., and Kaplan, J-C. *Proc. Natl. Acad. Sci. U.S.A.* **77**, 1917-1921 (1980)

168. Simondsen, R. P., Weber, P. C., Salemme, F. R., and Tollin, G. *Biochem.* **21**, 6366-6375 (1982)

169. Marcus, R. A. *Annu. Rev. Phys. Chem.* **15**, 155-172 (1964)

170. Poulos, T. L., Freer, S. T., Alden, R. A., Edwards, S. L., Skoglund, U., Takio, K., Erikkson, B., Xuong, Ng-h., Yonetani, T., and Kraut, J. *J. Biol. Chem.* **255**, 575-580 (1980)

171. Swanson, R., Trus, B. L., Mandel, N., Mandel, G., Kallai, O. B., and Dickerson, R. E. *J. Biol. Chem.* **252**, 759-775 (1977)

172. Takano, T. and Dickerson, R. E. *Proc. Natl. Acad. Sci. U.S.A.* **77**, 6371-6375 (1980)

173. Salemme, F. R. *Ann. Rev. Biochem.* **46**, 299-329 (1977)

174. Dickerson, R. E. and Timkovich, R. in *The Enzymes*, (P. D. Boyer,ed.) pp. 397-547, Academic Press, New York, 1975

175. The potential importance of Ile 81 in cytochrome *c* was brought to our attention by Prof. Richard Dickerson and we thank him for this insight.

176. Gupta, R. K. and Yonetani, T. *Biochim. Biophys. Acta.* **292**, 502-508 (1973)

177. Poulos, T. L., Freer, S. T., Alden, R. A., Edwards, S. L., Skoglund, U., Takio, K., Xuong, Ng.-h., Yonetani, T., and Kraut, J. in *Oxidases and Related Redox Systems* pp. 639-652, Pergamon Press, 1982

178. Bisson, R. and Capaldi, R. A. *J. Biol. Chem.* **256**, 4362-4367 (1981)

179. Bechtold, R. and Bosshard, H. R., personal communication.

180. Faraggi, M. and Pecht, I. *J. Biol. Chem.* **248**, 3146-3149 (1973)

181. Mitra, S. and Bersohn, R. *Proc. Natl. Acad. Sci. U.S.A.* **79**, 6907-6811 (1982)

182. Adman, E. T., Stenkamp, R. E., Sieker, L. C., and Jensen, L. H. *J. Mol. Biol.* **123**, 35-47 (1978)

183. Winfield, M. E. *J. Mol. Biol.* **12**, 600-611 (1965)

184. Dickerson, R. E., Takano, T., Kallai, O. B., and Samson, L. in *Structure and Function of Oxidation Reduction Enzymes* pp. 69, Pergamon Press, Oxford, 1972

185. Katon, J. E. ed. in *Organic Semiconducting Polymers*, Marcel Dekker,Inc., New York, 1968

186. Kepler, R. G. *Physiol. Rev.* **119**, 1226-1229 (1960)

187. Ward, R. L. and Weisman, S. K. *J. Amer. Chem. Soc.* **69**, 2086-2090 (1956)

188. McConnell, H. H. *J. Chem. Phys.* **35**, 508-515 (1961)

189. Reynolds, W. L. and Lumry, R. W. in *Mechanisms of Electron Transfer*, Roland Press, New York, 1966

190. Horne, R. A. *J. Inorg. Nucl. Chem.* **25**, 1139-1146 (1963)

191. Asakura, T. and Yonetani, T. *J. Biol. Chem.* **244**, 4573-4579 (1969)

192. Mochan, E. *Biochim. Biophys. Acta.* **216**, 80-95 (1970)

193. Cardew, M. H. and Eley, D. D. *Discuss. Faraday Soc.* **27**, 115-128 (1959)

194. Eley, D. D. and Spivey, D. I. *Trans. Faraday Soc.* **56**, 1432-1442 (1960)

195. Eley, D. D., Lockhart, N. C., and Richardson, C. N. *J. Bioenerg. Biomembr.* **9**, 289-301 (1977)

196. Rosenberg, B. and Postow, E. *Ann. N. Y. Acad. Sci.* **158**, 161-190 (1969)

197. Kertesz, M., Koller, J., and Azman, A *Phys. Rev. B.* **18**, 5649-5656 (1978)

198. Duke, C. and Schien, L. *Physics Today* **33**, 42-48 (1980)

199. Lewis, T. J. *Phys. Med. Biol.* **27**, 335-352 (1982)

200. DeVault, D. and Chance, B. *Biophys. J.* **6**, 824-847 (1966)

201. McElroy, J. D., Mauzerall, D. C., and Feher, G. *Biochim. Biophys. Acta.* **333**, 261-277 (1974)

202. Hopfield, J. J. *Proc. Natl. Acad. Sci. U.S.A.* **71**, 3640-3644 (1974)

203. Mauk, A. G., Scott, R. A., and Gray, H. *J. Amer. Chem. Soc.* **102**, 4360-4363 (1980)

204. Margalit, R., Pecht, I., and Gray, H. *J. Amer. Chem. Soc.* **105**, 301-302 (1983)

205. Ewall, R. X. and Barnett, L. E. *J. Amer. Chem. Soc.* **96**, 940-942 (1974)

206. Moore, G. R. and Williams, R. J. P. *Coord. Chem. Rev.* **18**, 125-197 (1976)

207. Vallee, B. L. and Williams, R. J. P. *Proc. Natl. Acad. Sci. U.S.A.* **59**, 498-505 (1968)

208. McKellar, J. R., Weightman, J. A., and Williams, R. J. P. *Discuss. Faraday Soc.* **51**, 176-182 (1971)

209. Maltempo, M. M., Ohlsson, P.-I., Paul, K.-G., and Ehrenberg, A. *Biochem.* **18**, 2935 (1979)

210. Maltempo, M. M., Moss, T. H., and Cusanovich, M. A. *Biochim. Biophys. Acta* **350**, 304 (1974)

211. Hintz, M. J. and Peterson, J. A. *J. Biol. Chem.* **256**, 6721-6728 (1981)

212. Osherofff, N., Borden, D., Koppenol, W. H., and Margoliash, E. *J. Biol. Chem.* **255**, 1689-1697 (1980)

213. Yonetani, T., Yamamoto, H., Erman, J. E., Leigh, J. S., and Reed, G. H. *J. Biol. Chem.* **247**, 2447-2455 (1972)

214. Poulos, T. L. and Edwards, S. L., Unpublished results.

215. Edwards, S. L., D. Phil. Thesis, Univ. California, San Diego, 1982

MOLECULAR ENZYMOLOGY OF
SELENO-GLUTATHIONE
PEROXIDASE

Rudolf Ladenstein

Abteilung Strukturforschung II
Max-Planck-Institut fuer Biochemie
Martinsreid, West Germany

ABSTRACT

The present knowledge on the molecular enzymology of seleno-GSH peroxidase is reviewed with emphasis put on the description of structure-function relationships. GSH peroxidase had been isolated from several sources. The bovine erythrocyte enzyme could be crystallized from phosphate buffer. The crystallographic refinement of the structure, determined by multiple isomorphous replacement led to new conclusions concerning the catalytic function and allowed the proposal of a tentative sequence which is in surprisingly good agreement with a partial sequence determined independently by chemical analysis. The enzyme is highly specific for GSH but reacts with a variety of hydroperoxides. A reasonable model, describing the binding interactions of the GSH molecule at the active site is presented. In accordance with the assumption that a selenol group is reversibly oxidized during catalysis, ping-pong kinetics are observed. Substrate and inhibitor binding studies are interpreted by an apparent half-site-reactivity. In conclusion, a general picture of a reaction mechanism which is in agreement with functional and structural data is proposed and unresolved problems concerning the catalytic function are discussed.

INTRODUCTION

The redox chemistry of oxygen is extensive and many of the oxygen derivatives are highly reactive and toxic to living organisms (Fig. 1).

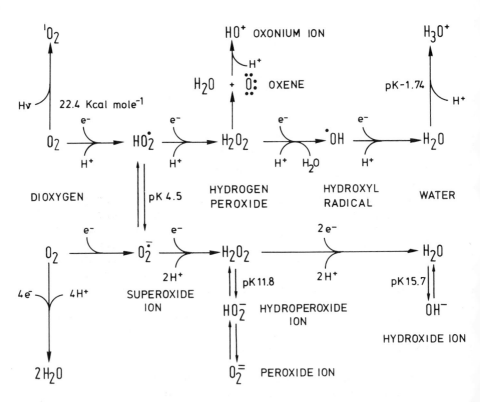

Fig. 1:   Pathways of oxygen chemistry

Nevertheless, nature encounters and frequently uses many of the intermediates of oxygen metabolism. In the case of the highly reactive species such as singlet oxygen and the hydroxyl radical their "indiscriminant" reactivity is used for the ultimate destruction of invading organisms and xenobiotics in the mammalian polymorphonuclear leukocyte [1], where myeloperoxidase plays a major role. Similarly, nature uses both heme and non-heme proteins in the manipulation of other oxygen derivatives.

Hydrogen peroxide is a powerful oxidizing agent and if left unattended in a living cell would soon create havoc. Life has devised at least four ways to handle this toxic yet necessary molecule in distinct compartments of the cell (Fig. 2):

(1)   The first line of defense seems to be governed by superoxide dismutase catalyzing the reaction

$$O_2^- \longrightarrow O_2 + H_2O_2$$

(2)   Seleno-GSH peroxidase and the heme protein catalase will, at essentially diffusion controlled rates, decompose H_2O_2 to oxygen and water. In the liver, for instance, the catalytic reaction is restricted to the peroxisomal space, whereas GSH peroxidase fulfills its task in the cytosolic and mitochondrial spaces.

(3)   On the other hand heme peroxidases use the oxidizing power of peroxide to bring about selective one-electron oxidations of organic substrates, usually phenols and amino-phenols.

(4)   In the last line of defense, which is required only if a polyunsaturated membrane lipid has been peroxidized, GSH peroxidase is the sole enzyme preventing further propagation of a radical chain reaction that leads to lipid peroxidation and severe disturbances of membrane function.

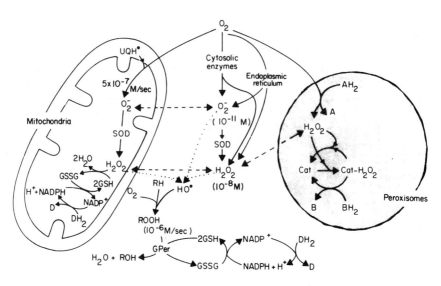

Fig. 2:   Sources and sinks for oxygen reduction
          products in the mitochondrial, cytosolic and
          peroxisomal spaces. (Adapted with permission
          from [60]). Abbreviations: UQH˙, ubiquinone
          radical, SOD, superoxide dismutase, Cat,
          catalase, GPer, glutathione peroxidase; the
          concentrations and formation rates of some
          metabolites are indicated.

Much new information has accumulated in the last few
years and several nutritionally related disorders as
well as diseases of mammals can now be attributed to
their inability to synthesize the specific seleno-
protein, glutathione peroxidase (E.C. 1.11.1.9). The
enzyme has been discovered in 1957 by Mills [2] as an
erythrocyte protein, which could protect hemoglobin from
oxidative destruction by catalyzing the reaction

$$2\ GSH + H_2O_2 \longrightarrow GSSG + H_2O$$

In 1968, a mitochondrial protein which was able to
prevent the GSH-induced swelling of mitochondria and
hence was named contraction factor could be identified
by GSH peroxidase [3]. It could be demonstrated that the

peroxidase exhibited a quite different, but nevertheless protective function in these organelles. Roughly at the same time, the enzyme was found to be capable of catalyzing the reduction of organic hydroperoxides, formed from unsaturated fatty acids [4,5]. Therefore the above reaction may be generalized to

$$2 \text{ GSH} + \text{ROOH} \longrightarrow \text{GSSG} + \text{ROH} + \text{H}_2\text{O}$$

introducing a broader role for the peroxidase in detoxication and protection of biomembranes. In 1973, the search for the catalytically active moiety of GSH peroxidase and the trace element research on the biological functions of selenium met at a common basis with the finding that GSH peroxidase contained selenium, bound in stoichiometric amounts [6,7]. Another selenium-independent GSH peroxidase activity was discovered in 1976 [8]. Unlike the selenium-dependent enzyme, the 47000-dalton protein exhibits negligible activity with hydrogen peroxide as substrate but seems to be rather active with organic hydroperoxides [9]. The presence of this enzyme in many tissues and its relation to GSH-S- Transferase B requires reevaluation of the role of GSH peroxidases in hydroperoxide metabolism and complicates the understanding of the metabolic interplay of detoxifying enzymes.

By studying the isolated perfused liver, the functions of the enzyme in this organ and its subcellular compartments could be evaluated in situ. As a consequence of the rapid GSH peroxidase reaction and the slow re-reduction of GSSG by the flavoenzyme GSSG reductase, GSSG is released from the cell. GSSG efflux could be shown by several independent groups [10, 11, 12]. If organic hydroperoxides, which cannot be decomposed by catalase, are infused into the isolated rat liver a proportional amount of extra GSSG is released that represents about 3% of the actual rate of the GSH peroxidase reaction [13]. Several similar effects have been observed during hyperoxic oxidative stress [12], intracellular action of drugs [14] and after intracellular generation of hydrogen peroxide [15]. By evaluating the ethane exhalation of selenium and vitamin E-deficient rats as a measure of in vivo

lipid peroxidation it could be shown that this event
occurs in vivo and may be responsible for oxidative
tissue damage. The effect can be considerably reduced by
a normal GSH peroxidase activity and by normal levels of
vitamin E [16]. These studies demonstrate convincingly
by noninvasive techniques that GSH peroxidase is
effectively decomposing H_2O_2 and organic hydroperoxides
within living cells.

The biological function of seleno-GSH peroxidase
has been treated comprehensively in several excellent
reviews. For more information concerning the relevance
of this enzyme in hydroperoxide metabolism, the reader
is referred to the following review articles: [17,18,
19].

ENZYMOLOGICAL DATA

Purification and Crystallization

GSH peroxidase is usually present in very low concentrations within the cell. It constitutes approximately 0.01 % of the total protein of bovine erythrocytes
[20] and roughly 0.1 % of the extractable rat liver
protein [21]. Rapid techniques and large-scale equipment
are needed for a successful isolation since the
peroxidase tends to become unstable upon purification.
Several groups have purified Se-GSH-peroxidase from
different sources [21, 22, 23, 24, 25, 26, 27]. The
Tuebingen group seems to be most successful in obtaining
very high specific activities and considerable yields:
the currently used isolation strategy there includes
hollow-fiber dialysis of the hemolysate, ion-exchange
chromatography on DEAE sephadex, hydrophobic
chromatography on phenyl-sepharose, gel filtration on
Sephacryl S-300 and hydroxyapatite chromatography [28].
The procedure yields about 200 mg of electrophoretically
homogeneous enzyme with a specific activity of up to 600
U/mg from 180 liters of bovine blood.

For the determination of the enzymatic activity of
GSH peroxidase two assay methods have been developed and
are currently used:

(a) A direct fixed-time assay which measures the remaining GSH either polarographically by the p-chloromercuribenzoic acid method, or by the dithionitrobenzoic acid method. Due to many possibilities of interference with components occurring in biological material this assay can only be used with purified enzyme [29, 30, 31].

(b) In a coupled test procedure the product GSSG is regenerated enzymatically by GSG reductase and the corresponding decrease in NADPH concentration is observed spectrophotometrically. The coupled test is suitable for the determination of GSH peroxidase activity in biological samples, but does not allow kinetic studies. [21, 23, 32, 33]

Seleno-GSH-peroxidase has been crystallized for the first time in 1973 from 1.2 M phosphate buffer, pH = 7.0 at 4°C [34]. Rather flat, plate-like crystals not suitable for X-ray analysis were obtained. Later the growth of GSH peroxidase crystals from 1.7 M ammonium sulphate, pH = 6.4, 20°C has been described [35]. These crystals, however, could not be obtained reproducibly. Prismatic, plate-like crystals, suitable for X-ray data collection could be obtained by the vapour diffusion method at 4°C [36]. Prior to crystallization, the enzyme was converted to a defined oxidized state by stepwise dialysis against 10 mM GSH, then 5 mM H_2O_2 and finally 5 mM GSSG in 0.5 M phosphate buffer, pH = 7.0 (Fig. 3).

The best crystals could be grown from 1.5 M phosphate buffer of the same pH. The final protein concentration was about 5 mg/ml. For X-ray analysis the crystals were transferred to a storage solution consisting of 2.5 M phosphate buffer, pH = 7.0, 5 mM GSSG, 1 mM EDTA, 1 mM sodium azide. In Table 1 the crystallographic data of bovine erythrocyte GSH peroxidase together with some important molecular data are summarized.

Substrate Specificity and GSH Binding

GSH peroxidase catalyzes the reduction of a broad variety of hydroperoxides to the corresponding alcohols.

Fig. 3:   Monoclinic GSH peroxidase crystals grown from 1.5 M phosphate buffer, pH = 7.0; space group C2, dimensions 0.5 x 0.3 x 0.1 mm.

Table 1

Molecular and Crystallographic Properties of Seleno-
GSH-Peroxidase from Bovine Erythrocytes [20,35,36]

Molecular weight	M = 84000 (tetramer)
Subunits	M = 21000, identical
Monomer radius	R = 19 $\overset{o}{A}$
Molecular symmetry	222
Isoelectric point	pJ $\approx$ 5.8
pH optimum	pH = 8.8
Temperature optimum	T = 42°C
Prosthetic groups	1 selenocysteine/monomer
Crystal space group	C2, monoclinic
Unit cell constants	a = 90.4 $\overset{o}{A}$, b = 109.5 $\overset{o}{A}$, c = 58.2 $\overset{o}{A}$, = 99° + 15$^{\prime}$
Unit cell volume	V = 569 000 $\overset{o}{A}{}^3$
Subunits per asymmetric unit	dimer, M = 42000
Matthew's parameter	V_m = 3.39 $\overset{o}{A}{}^3$/Dalton
Resolution	measurable reflexions to spacings of 2.0 $\overset{o}{A}$
Crystal solvent content	64 %

The only efficient physiological donor substrate appears to be GSH. The hydroperoxides accepted as substrates include H_2O_2, ethyl hydroperoxide, t-butyl-hydroperoxide, cumene hydroperoxide, thymine hydroperoxide, hydroperoxides of unsaturated fatty acids and the corresponding esters, hydroperoxides of steroids and nucleic acids and prostaglandin G_2, the primary intermediate of prostaglandin biosynthesis (for review see [19,37]). It is thus tempting to speculate that the enzyme reacts unspecifically with every hydroperoxide, unless the reaction with the hydroperoxy group is not sterically hindered. Dialkyl peroxides as well as cyclic peroxides are apparently not metabolized.

Specificity studies with various thiol compounds indicate that both carboxyl groups of the GSH molecule contribute to substrate binding. An enormous decrease of the enzymatic activity is observed if the γ-Glu residue of GSH is substituted by a β-Asp or N-acetyl residue or if the glycine residue is replaced by a methoxy or amide group (Table 2).

For the bovine enzyme the existence of a typical ES-complex in presence of excess GSH (enzyme/GSH = 1:10) could not be demonstrated by binding studies [39]. In another attempt substrate binding to the oxidized enzyme has been studied by co-chromatography with labeled GSH [40]. The experiment was performed at low GSH concentration with ^3H - GSH in an equimolar ratio with respect to the number of the active sites of the enzyme. Although this binding experiment was carried out under non-equilibrium conditions, a considerable amount of the labeled GSH, about 10% of the total, remained bound to the enzyme and could, in a subsequent step, be removed by dialysis.

The experiment demonstrates that GSH is bound noncovalently to the peroxidase, presumably by a rather weak binding interaction, which seems to be more an orientation process through electrostatic forces, inducing the proper binding geometry of the GSH molecule at the active site, than the formation of a stable ES-complex in the usual sense.

Considering the covalent binding of GSH to the peroxidase, the observed binding stoichiometries appear

Table 2

Characteristic Examples, Showing the Dependence of the
Activity of Se-GSH-Peroxidase on Structural Variations
of the GSH Molecule. (Data are taken from [38].)

GSH analogue		Enzymatic activity (%)
γ-Glu-Cys-Gly	(GSH)	100.0
β-Asp-Cys-Gly	Variations of	7.6
Cys-Gly	γ-Glu	6.8
N-Ac- Cys-Gly		2.7
γ-Glu-Cys-OMe	Variations	26.0
γ-Glu-Cys-NH$_2$	of Gly	1.4

confusing: Under oxidizing conditions in solution GSH
peroxidase seems to contain one covalently bound GSH
molecule per subunit, presumably via a selenosulfide
linkage (E-Se-SG) [41]. In a labeling experiment with
crystalline GSH peroxidase binding of 2 moles GSH/mole
tetramer was observed reproducibly [40]. Since extensive
washing of the crystals in phosphate buffer prior to the
radioactivity measurements did not remove the bound
radioactivity, it was concluded that GSH has formed
covalent bonds with the crystalline enzyme.

The crystallographic analysis of GSH binding [40]
by difference Fourier methods could be carried out only
under great difficulties, since the observed binding
stoichiometry in the crystalline state would allow only
a maximum overall occupancy of 0.5 GSH per binding site.
The clearest, although ill-defined, density features
supporting binding interaction of GSH at the
selenocysteine residue was observed in a difference
electron density map between GSH reduced crystals and

the oxidized enzyme. Since the resulting difference
density appeared discontinuous and poorly defined, the
structure of the ES-complex was not refined crystallographically, but a hypothetical model of GSH binding,
based on an interpretation of the difference density
giving the most probable fit, could be proposed (Fig. 4).
During binding interaction the γ-Glu carboxyl-group of
the GSH molecule is presumably fixed by a salt-bridge to
Arg (167), whereas the N-terminal amino group could form
a hydrogen bond to Glu (130). The C-terminal
Gly-carboxyl group which, due to the steric conditions,
should point into the solvent space, seems to form an
additional salt-bridge with Arg (140) which is a part of
helix α_1. The distances between these functional

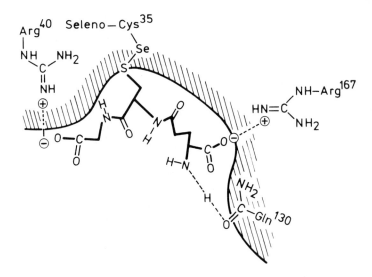

Fig. 4:   Proposed model of GSH binding interaction at
           the active sites of GSH peroxidase. The
           γ-Glu-carboxyl group and the C-terminal
           Gly-Carboxyl of the GSH molecule are fixed by
           salt bridges to Arg 167 and Arg 40,
           respectively. The N-terminal amino group could
           form a hydrogen bond to Gln 130. (Adapted with
           permission from [40]).

residues of the peroxidase and their presumed counterparts on the GSH molecule are in favour of the hypothetical binding model.

The Kinetic Mechanism

The dependence of the initial rate on the concentrations of the substrates can be described by a differential equation analogous to that developed for other peroxidases [42]:

$$\frac{[E_o]}{[R\overset{\bullet}{O}OH]} = \frac{\Phi_1}{[ROOH]} + \frac{\Phi_2}{[GSH]} \tag{1}$$

The kinetic constants Φ_1, Φ_2 are defined by the following relations

$$\Phi_1 = \frac{1}{k_{+1}} \quad , \quad \Phi_2 = \frac{1}{k_{+2}} + \frac{1}{k_{+3}} \tag{2}$$

where k_{+1} describes the reaction of the reduced enzyme with the hydroperoxide, k_{+2} and k_{+3} represent the rate constants of the reaction of the oxidized enzyme species with GSH and $[E_o]$ stands for the total enzyme concentration. Equation (1) describes a ping-pong mechanism, which may be written in Cleland's terms as follows:

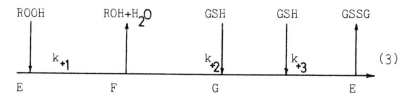

The absence of a kinetic constant Φ_o implies that no enzyme-substrate complexes are built, or at least, that the formation of complexes E·GSH does not influence the initial velocity, since these complexes do not accummulate even at high GSH levels. The kinetic data for GSH peroxidase from bovine blood [43, 44] and rat liver [45] can be satisfactorily described by equation (1) and the reaction scheme (3). The kinetic constants are summarized in table 3. The first order rate constant k_{+1}, of the oxidation of the reduced enzyme species by the hydroperoxide substrate is in the order of 10^{8} · M^{-1}·s^{-1} (pH = 7.0, $37^{\circ}C$) and reflects a diffusion-controlled reaction. This is in accord with the exposure of the enzyme-bound selenium on the surface of the molecule (see below). The mechanism is best interpreted by the assumption that the enzyme goes through consecutive steps of reduction and oxidation during catalytic

Table 3

Kinetic Constants for Glutathione Peroxidase from Bovine Erythrocytes (Data taken from [43])

Substrate	pH	$\Phi_1 \, 10^{-8}$ M · s	$\Phi_2 \, 10^{-6}$ M · s
Hydrogen peroxide	6.7	1.70	2.19
Ethyl hydroperoxide	6.7	3.3	2.24
Cumene hydroperoxide	6.7	7.8	2.24
t-Butyl hydroperoxide	6.7	13.5	2.24

$K_m app(H_2O_2)$ = 0.008 mM ([GSH] = 2.5 mM)
$K_m app(GSH)$ = 0.130 mM ([H_2O_2] = 0.001 mM)

action. Certain characteristic features of this kinetic
mechanism however deserve mention:

(a) Within the considered concentration ranges the
    peroxidase shows no saturation with respect to GSH.
    hence only apparent K_m values can be given.

(b) The kinetic mechanism implies the existence of
    three enzyme species, exhibiting different redox
    states (Fig. 5).

Inhibition Studies

Oxidized GSH peroxidase can be inactivated and
depleted of its selenium after prolonged exposure to
high concentrations of potassium cyanide in solution as
well as in the crystalline state [36,40,41]. Selenium
depletion led to the identification of four selenium
sites in the tetrameric enzyme as the most prominent
features in a difference electron density map calculated
between selenium-free and native enzyme crystals. By
difference Fourier analysis of the deseleno enzyme [40]
strong evidence had been provided that the reaction of
the selenocysteine residue with cyanide proceeds via a
displacement reaction which could described by

$$R-CH_2-Se^- + CN^- \longrightarrow R-CH_2-CN + Se^{2-}$$

Inspection of a difference Fourier map calculated with
coefficients $(|F_o (DESE)| - |F_c (DESE)|) \cdot exp i \alpha_c (DESE)$
revealed a residual density maximum adjacent to C_β of
the selenocysteine residue which can be readily
interpreted by a bound cyano group.

Se-GSH peroxidase can be inhibited readily by
treatment with iodoacetate [46] or chloroacetate [47] if
reduced by GSH or KBH_4. The stoichiometry of the binding
of labeled haloacetates is shown in table 4.

Curiously, iodoacetamide does not inactivate the
peroxidase. An explanation for the finding that 2 moles
of chloroacetate or iodoacetate are bound to the
tetrameric enzyme concomitant with complete inactivation

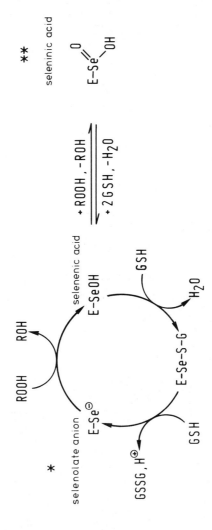

Fig. 5: Hypothetical mechanism for the hydroperoxide reduction
catalyzed by GSH peroxidase.

Table 4

Binding Stoichiometry of [^{14}C] Labeled Haloacetates
to Native and Cyanide-Treated GSH Peroxidase
of the Reduced Form.
(Data are Taken from [47].)

Alkylating reagent	Treatment	Reagent bound per tetramer	Se content per tetramer
Jodoacetate	-	2.1	3.5
Jodoacetamide	-	0.1	4
Chloroacetate	-	2.2	4
Chloroacetate	Cyanide	0.1	0.02

will be given below. The control experiment with cyanide
treated enzyme showed that the presence of selenium is
required for chloroacetate binding. The binding energy
of the Se 3d electrons measured by X-ray photoelectron
spectroscopy indicates some change in the chemical
bonding of the Se atom and with 56.8 eV a position
between the values for reduced (54.1 eV) and oxidized
(58.0 eV) peroxidase [47, 48]. These data seem to allow
the statement that the active-site selenium can be
chemically modified by treatment with haloacetate
compounds.

In an attempt to titrate the active sites of
reduced GSH peroxidase the inhibition by iodoacetate has
been used as a tool to evaluate the number of active
sites [34, 46]. Stepwise addition of t-butyl
hydroperoxide to GSH-reduced enzyme resulted in a
gradual decrease of the inhibition by iodoacetate until
full protection of the enzyme was achieved. This point
was reached at a molar ratio of hydroperoxide/
tetrameric enzyme of 2.2, but the authors discussed 4

active sites, assuming reoxidation of the enzyme prior
to titration.

THE THREE-DIMENSIONAL STRUCTURE

The Symmetry of the oligomeric Molecule

Tetrameric GSH peroxidase consists of four chemically indistinguishable subunits. The asymmetric unit of
the monoclinic crystal cell is occupied by a dimer [36]
and the monomers are related by non-crystallographic
symmetry, which could be established by analysis of
Patterson autocorrelation functions [36, 49]. Fig. 6
shows a stereogram plot of the rotation function for
twofold symmetry axes.

Apart from the central peak, reflecting the crystallographic twofold b-axis, the highest features within
the stereogram are two peaks (height = 65.0 arbitrary
units) indicating local twofold axes perpendicular to
each other and to the crystallographic dyad.
Consequently, the tetrameric molecules show [222]-sym-
metry proving the subunits to be identical or at least
very similar. A molecular coordinate system relative to
the crystal axes can be defined by the orthogonal axes
given in table 5.

The molecular symmetry and the packing of GSH peroxidase tetramers in the monoclinic crystal cell is shown
schematically by Figure 7. The tetrameric arrangement of
the subunits is rather flat, very similar to an almost
planar arrangement of four monomers.

The data obtained by crosslinking studies in solution with bifunctional reagents [50] fit best to the
assumption of a square model composed of isologous
dimers, showing D_2 symmetry in total. Hence, two
completely different methods led to very similar conclusions on the quaternary structure of GSH peroxidase.
A linear correlation established between surface area
accessible to solvent and hydrophobic free energy [51]
suggests that the reduction of accessible surface area
occurring upon subunit association is the main source of
free energy ($\Delta G_{transfer}$) stabilizing protein-protein

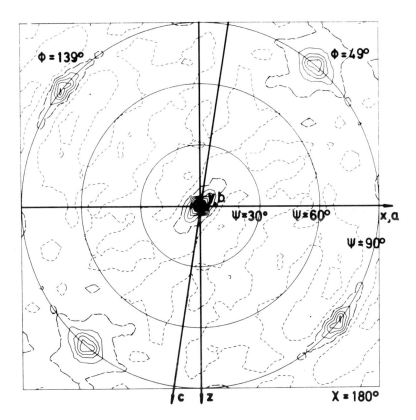

Fig. 6:   Stereogram plot for 2-fold symmetry axes (X = 180°) derived from correlation calculations in Patterson space. Polar angle definition according to [49]. Height of origin peak, 100 arbitrary units, mean value   of   rotation function (— — —) 25.5, contour   line separation, 10, values above mean   value (————————), (Adapted with permission from [36].)

Table 5

Refined Polar Coordinates of the Twofold Symmetry Axes
Relating the Subunits of a GSH Peroxidase Tetramer
(Polar Angle Definition According to [49])

Axes	polar	angles (^{0})	type
P	90.0	138.3	local
Q	90.0	48.3	local
R	0.0	0.0	crystallographic

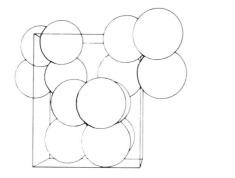 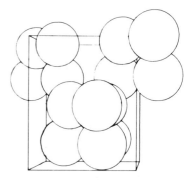

Fig. 7:   Symmetry and   molecular   packing   of GSH
            peroxidase tetramers within the crystal cell,
            view along, z axis   (stereo   pair). (Adapted
            with permission from [36].)

interactions. In view of the similar magnitudes of the monomer contact areas buried on association ($1427 \, \text{Å}^2$ at the local axis and $1472 \, \text{Å}^2$ at the crystallographic axis, respectively) roughly identical values of the individual association constants between the GSH peroxidase monomers had been predicted.[40]

The Folding of the Peptide Chain

A schematic drawing of the secondary structure elements of a GSH peroxidase subunit is shown in Fig. 8. A subunit is built up from a central core of two parallel (β_1, β_2) and two antiparallel (β_3, β_4) strands of pleated sheet, surrounded by four α-helices. This pleated sheet structure has a right-handed twist of about 45 degrees. The helices α_1, α_2 and α_4 are placed on one side of the β-structure and helix α_3 on the other side. Twelve clearly defined β-turns of types I, II, III and III' [52] as well as three one-turn 3_{10}-helices could be recognized in the monomer structure.

Figure 9 shows a stereo diagram of the C_α backbone of the tetrameric molecule. The distances of the selenium atoms across the molecular axes P, Q, R are $20.7 \, \text{Å}$, $39.3 \, \text{Å}$, and $36.3 \, \text{Å}$, respectively.

Unfortunately, the amino acid sequence of the bovine blood enzyme has not been determined by chemical analysis. Although the crystallographic refinement has been brought to an end, there is uncertainty in the identity of some residues, as the identification is based primarily on the shape of the electron density. The amino acid sequence of the final model given in Table 6 must still be regarded as an approximation, but this form of description seems to be more informative than a shape classification, even if tentative. The proposed sequence is in good agreement with an amino acid analysis of bovine erythrocyte GSH peroxidase [34]. In addition, the high resolution and the well-refined analysis provides confidence.

Recently an important progress in the sequence analysis of rat liver GSH peroxidase could be achieved [56]. A peptide consisting of 16 amino acid residues which includes the active site selenocysteine residue was

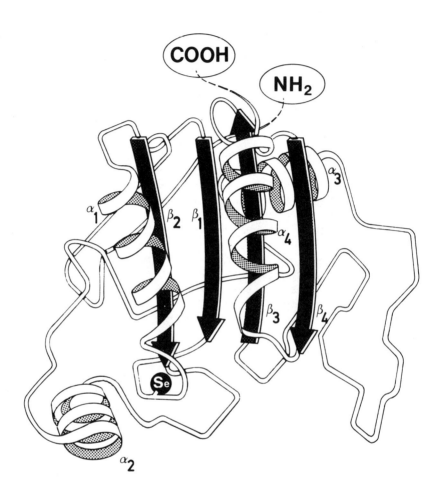

Fig. 8:   Schematic drawing of the folding pattern of a
          GSH peroxidase monomer; α_i and β_i indicate
          α-helices and β-strands, respectively; Se
          shows the position of the selenocysteine
          residue (SeCys 35) (Adapted with permission
          from [40]).

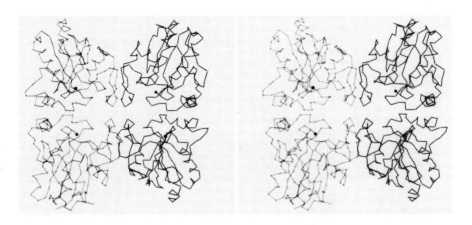

Fig. 9: C_α backbone of a GSH peroxidase tetramer (Stereo pair); O positions of C_α atoms; ● positions of selenium atoms. (Adapted with permission from [36]).

Table 6

Tentative Amino Acid Sequence of a Glutathione
Peroxidase Monomer from Bovine Erythrocytes
(Determined by Interpretation of the
Electron Density Distribution)

```
Ala Thr Val Tyr Ala Phe Ser Ala Arg Pro Leu Gly Gly
Gly Pro Phe Ser Leu Ser Ala Leu Arg Gly Lys Val Leu
Leu Ile Gln Asn Val Ala Ser Leu Cso Gly Thr Thr Thr
Arg Asn Tyr Ser Gln Gln Asn Asp Leu Gln Gln Arg Leu
Gly Pro Arg Gly Leu Val Val Leu Gly Phe Pro Cys Asn
Gln Phe Gly Met Gln Gln Asn Ala Lys Asn Gln Glu Ile
Leu Asn His Leu Gln Tyr Val Arg Cpr Gly Gly Gly Phe
Gln Pro Asn Phe Leu Leu Phe Gln Lys Met Lys Val Asn
Gly Ala Ser Ala His Pro Leu Phe Ala Phe Leu Arg Lys
Val Leu Pro Val Pro Ser Thr Asp Ala Thr Ala Leu Gln
Thr Asn Pro Gln Phe Ile Thr Trp Ser Cpr Val Gln Arg
Asp Asp Val Ser Trp Asn Phe His Lys Phe Leu Val Gly
Pro Asn Gly Thr Pro Thr Arg Arg Tyr Ser Arg Arg Phe
Leu Thr Ile Asn Ile Glu Pro Asn Ile Ser Thr Leu Leu-COOH
```

Cso = seleninic acid
Cpr = cis-proline
----

isolated from a tryptic digest by reverse-phase HPLC and gel filtration. The amino acid sequence of the first 46 N-terminal residues could be obtained from the overlapping sequences of this tryptic selenopeptide and by sequencing of the intact subunit from its N-terminal end. In table 7 the residues determined by sequencing the N-terminus (N) and the tryptic selenopeptide (T) are shown.

A surprising agreement between the sequences determined independently by chemical analysis of the rat liver enzyme and by crystallographic refinement of the structure of the bovine erythrocyte enzyme can be stated. Considering the tryptic selenopeptide (T), the correlation is 100%. A correlation of 75% could be calculated for the N-terminal peptide (residues Gly 1 - Lys 30) by taking into account that the first five N-terminal residues might be flexible and hence are not defined in the electron density distribution.

On the other hand the sequence of the rat liver enzyme must not necessarily be identical with the sequence

Table 7

Amino Acid Sequence around the Active Site
Selenocysteine of Rat Liver GSH Peroxidase.
(Data are taken from [56].)

```
-------------------------------------------------------------
               5                        10
H₂N-Gly-Ala-Val-Ala-Glu-Ser-Thr-Val-Tyr-Ala-Phe-Ser-
├──────────────────────────── N ─────────────────────

              15              20                    15
Ala-Arg-Pro-Leu-Ala-Gly-Gly-Glu-Pro-Val-Ser-Leu-Gly-Ser-
─────────────────────────── N ───────────────────────

              30              35                    40
Leu-Arg-Gly-Lys-Val-Leu-Leu-Ile-Glu-Asn-Val-Ala-Ser-Leu-
─────────────────────────── N ───────────────────────
             ├─────────────── T ─────────────────────
              45
SeCys-Gly-Thr-Thr-Thr-Arg-COOH
───── N ────┤
─────────────────────────── T ──┤
-------------------------------------------------------------
```

of the bovine erythrocyte enzyme. In the view of these
correlations, crystallographic refinement of protein
structures with unknown sequences must be regarded as a
potent method to obtain sequence information from X-ray
crystallographic measurements.

The chemically determined sequence was analysed by
computer search for homology with other proteins, but no
closely related sequences were found, although certain
parts of the GSH peroxidase fold are similar to the
structures of bacteriophage T4 thioredoxin [53] and
rhodanese [57].

An analysis of the atomic temperature factors allowed some conclusions on the mobility of the structure
[40]. Especially some strands forming β-sheets (β_1,β_4),
two of the α-helical structures (α_1, α_3) and certain
parts of the chain involved in subunit contacts show significantly lower mean square displacements than other
parts without well-defined secondary structure. It is
also evident that the mobility of the main chain in the
central regions of the α-helices is lower than at the
end. N- and C-terminal residues show high temperature
factors indicating that these parts are relatively
flexible. The highest values can be found at the
positions of a narrow loop (residues 13-15) near to the
N-terminus and some β-turns. These residues are exposed
to solvent and therefore are less constrained by packing
effects. The chain segments around the active site
(SeCys 35) are formed by a well-defined secondary
structure element, a $\beta\alpha\beta$-structure. They are involved in
hydrogen bond interaction and exhibit low mean square
displacements, causing a relatively rigid arrangement of
the peptide chains at the active center region.

A comparison of the GSH peroxidase subunit structure
with bacteriophage T4 Thioredoxin [53] revealed large
regions of structural resemblance. Thioredoxin shows
almost the same folding pattern as the core structure
of a GSH peroxidase subunit (see Figure 10),
four strands of pleated sheet and two adjacent
α-helices. In place of the reactive selenocysteine, T4
Thioredoxin contains a redox active disulfide at an
analogous position of the $\beta\alpha\beta$-unit. The striking
similarity of the active regions of these two redox

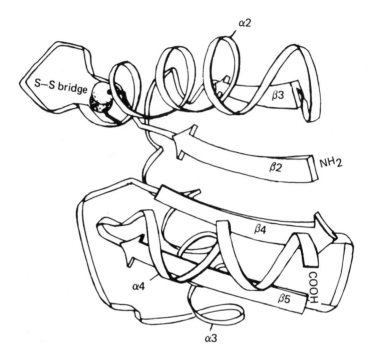

Fig. 10:   Folding of bacteriophage T4 Thioredoxin, α_i
           and β_i indicate α-helices and β-strands; -S-S
           shows the position of the redox active
           disulfide bridge. (Adapted with permission
           from [53]).

proteins may be the result of convergent evolution or
alternatively may indicate divergence from a common
ancestral redox molecule.

        Considering the T4 Thioredoxin fold as a
fundamental element of the GSH peroxidase monomer, it
seems reasonable to suggest that during protein
evolution two insertions into the backbone of a monomer
may have provided additional structure elements (two
loops formed by residues 67-73 and 124-140, helix α_2)
forming contact sites for protein-protein-interaction,
which have allowed the specific assembly of the
oligomeric enzyme molecule.

The Structure of the Active Site

The catalytically active selenocysteine residue 35 is located at the N-terminal end of helix α_1 which forms a $\beta\alpha\beta$-substructure together with the two adjacent parallel β-strands (β_1, β_2). Similar arrangements of secondary structure are frequently observed at active regions of enzymes that are folded into α- and β-structural elements. In general, $\beta\alpha\beta$ structures provide a geometrically favourable situation for the occurrence of a binding region near to the carboxy ends of two adjacent parallel β-strands [54]. Since the active site of GSH peroxidase can be localized in the general vicinity of SeCys 35, where the carboxy ends of two parallel strands (β_1, β_2) meet one another, it is tempting to speculate that this arrangement is of unique importance for catalysis. It has long been known that in the α-helix the alignment of the peptide dipoles parallel to the helix axis gives rise to a macrodipole of considerable strength. For points near to the helix N-terminus the effect of the dipole is equivalent to the effect of half of a positive unit charge [55]. Hence one could imagine that the electric field due to the dipole moment of helix α_1 will stabilize the active site selenolate and enhance its nucleophilic ractivity. The charged cosubstrate GSH, representing itself a dipole, may be oriented by this field prior to the electron transfer process.

The active sites of GSH peroxidase are localized in flat depressions on the molecular surface and are readily accessible. In a space-filling drawing of the molecule (Fig. 11) no prominent clefts or crevices are visible at the active centre regions. Exposure of the catalytically active seleno-cysteine residues at the molecular surface is consistent with the easy access of the substrates and thus the high reaction rates of the enzyme.

In the refined electron density map of the oxidized enzyme no covalent connections of the seleno cysteine side chains to ligands in its environment can be observed. Hence the active centres contain free selenocysteines or the corresponding oxidized derivatives, depending on the functional state of the enzyme. Figure 12 shows the chemical environment of the

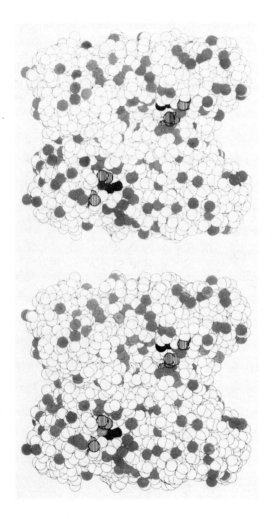

Fig. 11 : Space-filling drawing of a GSH peroxidase tetramer in the oxidized state (stereo pair). The oxygen atoms corresponding to the seleninic acid group of the enzyme are indicated by black spheres. Bound water molecules are represented by grey spheres and the residues presumably involved in GSH binding (Arg 40, Arg 167, Gln 130) are marked with small lines. (Adapted from [40].)

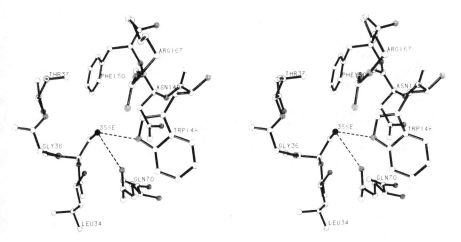

Fig. 12:   Chemical   environment   of   the   selenocysteine
            residue in the reduced enzyme (stereo pair);
            (●) position of the selenium atom; (----)
            hydrogen bonds to 148 Trp-N and 70 Gln-N.
            (Adapted with permission from [40]).

seleno cysteine residue in reduced GSH peroxidase. At
physiological pH-values a selenolate anion rather than an
undissociated free selenol seems to occur in the reduced
enzyme. Additional evidence is provided by difference
Fourier analysis between inhibited enzyme forms and the
reduced enzyme [40].

     The distances observed at a Vector General computer-
graphics system are consistent with hydrogen bond
formation of the selenolate anion to 148 Trp-N (d = 3.4
Å) and to 70 Gln-N (d = 3.3 Å), which would result in
some stabilization of the active site geometry (see
Table 8). A difference Fourier analysis between oxidized
and reduced GSH peroxidase has provided evidence [40]
for a seleninic acid derivative (E-SeOOH)   with
pyramidal arrangement of β-carbon , selenium and oxygen
atoms existing in the peroxide -oxidized enzyme form.
The seleninic acid derivative could be readily reduced
by GSH in the crystalline state to an enzyme form
containing the catalytically active selenol and could

Table 8

Coordination of the Active Site in Oxidized and
Reduced GSH Peroxidase (Data are Taken from [40])

Functional state	coordinated atom	coordination partner	distance [$\overset{O}{A}$]
oxidized	oxygen(1)	N-Trp(148)	3.2
		N-Gln(70)	3.0
		N-Gly(36), main chain	2.9
$E-S\overset{O}{\overset{\|}{e}}-OH$	oxygen(2)	O-Thr(37)	2.9
		N-Thr(37), main chain	2.9
		O-H_2O bound water	3.0
reduced	selenolate anion	N-Trp(148)	3.4
$E-Se^-$		N-Gln(70)	3.3

----

also be stabilized through coordination via hydrogen
bonds (Table 8).

The occurrence of selenocysteine residues in the
active centres of GSH peroxidase poses an interesting
question as to how this unusual amino acid is
incorporated into this enzyme. In most cases a specific
posttranslational modification of an existing amino acid
in a polypeptide chain seems to be responsible for the
occurrence of an abnormal amino acid in a protein
molecule. A selenocysteine may be formed by addition of
selenide or selenoglutathione (G-S-SeH) to a
dehydroalanine generated from a specific serine or
cysteine.

During the last year, however, recent experimental evidence has accumulated that rat liver contains a form of tRNA that is aminoacylated specifically with seleno-cysteine [58]. Although the existence of a specific selenocysteyl tRNA is consistent with the idea that selenocysteine might be incorporated intact into seleno proteins during protein synthesis, it does not conclusively prove this point. It remains to be determined if this type of pathway occurs in vivo and, if it does, to what extent it could account for the synthesis of known selenocysteine - containing proteins. These results are in contrast to reports of similar experiments in E. coli [59] where only insignificant differences were observed between the behavior of cysteyl and selenocysteyl tRNA.

Towards an Understanding of the Catalytic Process

In the view of the binding data obtained with different inhibitors and the reducing substrate half-site reactivity has been postulated for GSH peroxidase [40]. Half-site reactivity is related to negative cooperativity and indicates that constitutent protomers of an oligomeric enzyme do not behave as independent entities. This would give a functional meaning to oligomeric structures exhibiting classical Michaelian kinetics. Mechanisms describing the functional properties of half-site enzymes have been proposed [61] and assume a functional interrelationship between distinct active sites on identical monomers. It is concluded further that these oligomers behave as "polydimers" and, indeed, a polydimeric character appears to be a fairly general feature of enzymes with subunit structure.

The fact that in crystalline GSH peroxidase the dimers of an asymmetric unit exhibit local twofold symmetry suggests that small structural differences might exist between the monomers. It has been postulated that the tetramers might function as $(\alpha\beta)_2$ entities. Probably the dimeric structure loses its symmetry by binding of one GSH molecule.

There are several lines of indirect evidence which might indicate structural rearrangements upon GSH binding. A highly structured UV difference spectrum, and perturbations in the UV circular dichroism spectrum are observed under the influence of the reducing substrate [39]. A similar behavior could also be observed by difference Fourier analysis of reduced and oxidized enzyme forms [40]. Hence small and subtle conformational changes or side chain movements are likely to occur when the oxidation state of the peroxidase is changed. There is no crystallographic evidence however, indicating the occurrence of large conformational changes upon reduction by GSH.

The catalytic process can be described by the following hypothetical reaction mechanism, which is supported by a broad basis of experimental results:

(a) substrate specificity studies

(b) analysis of the kinetic mechanism

(c) inhibitor and substrate binding data

(d) biochemical investigations on the nature of the enzyme-bound selenium

(e) X-ray photoelectron spectroscopic studies

(f) X-ray structure analysis of reduced and oxidized enzyme forms.

The reduced form of the enzyme reacts with a hydroperoxide substrate in a bimolecular reaction. Probably this reaction proceeds without the formation of a specific enzyme - substrate complex. The reduced enzyme most likely contains a largely dissociated selenol group of SeCys 35, a selenolate anion ($E-Se^-$). This negatively charged anion can be stabilized by hydrogen bonding to N-Trp 148 or N-Gln 70. The interplay between SeCys 35 and Trp 148 seems to be important for catalysis, but is not completely understood at present. Several oxidized forms of the peroxidase may exist, depending on the relative concentrations of reducing and oxidizing substrates. Assuming physiological substrate concentrations (GSH in the mM range and hydroperoxides in the μM range) the main reaction would certainly shuttle between a selenolate anion and a selenenic acid derivative [E-SeOH] of the active site selenium. In the peroxide-oxidized enzyme (5 mM H_2O_2) however, a seleninic acid derivative (E-SeOOH) could be detected by X-ray crystallography [39]. But this form, although it could be reduced to the active site selenol in the crystalline state, is assumed to represent an unphysiological oxidized form of the peroxidase. A mixed selenosulphide (R-Se-S-R') can be excluded as an alternative, since no sulfur amino acid side chain in a distance from the active site compatible with the formation of covalent bonds could be detected in the molecule. The reduction of the oxidized enzyme is initiated by formation of an [E·GSH] complex. This complex is transformed into a mixed selenosulphide intermediate E-Se-SG. The formation of the complex [E·GSH] seems to be much slower than the intramolecular transformation into the covalent intermediate. In the last step of the mechanism the second GSH molecule restores the reduced enzyme form (Figure 5).

The catalytic events taking place at different active sites of a dimer are presumably controlled by negative cooperativity implicating that chemical signals, e.g. small conformational changes and/or proton displacements, are to be transferred across the twofold symmetry axis of the dimer. In this hypothetical model the active sites would react in a flip-flop type mechanism [61] influencing each other over a distance of roughly 20 Å. A likely candidate able to mediate a hydrogen bond displacement across the twofold axis is the couple of sidechains interpreted as His 81, His 281 residues. The N_ϵ atoms of these histidines are about 3.5 Å apart, allowing hydrogen bond interaction. If these residues are blocked by peroxodisulphate, presumably in a complex formation reaction, the enzyme loses its catalytic activity [40]. Modifications at the selenium sites other than seleninic acid formation could not be observed upon peroxodisulphate treatment. Conclusions on the mechanism and the involvement of additional charged side chains

must await the chemical analysis of the amino acid
sequence of the bovine erythrocyte enzyme.

UNRESOLVED PROBLEMS REMAIN

There is no doubt that GSH peroxidase is one of the
best-investigated enzymes. At least some aspects of its
biological role can no longer be questioned, despite the
emerging complexity of hydroperoxide metabolism. The
knowledge of its three-dimensional structure has brought
a deeper understanding of the structure - function
relationships and has led to the rejection of wrong
catalytic models. There exist, however, some unresolved
problems, concerning the catalytic function of this
enzyme, which might be difficult to solve and it may be
timely to add a few more questions:

(1)   Is the site responsible for hydroperoxide reduction
      identical with the selenium site ? One weak point
      of the proposed reaction sequence in general is,
      that the identity of the peroxide binding site and
      the selenium site up to now could not be proven
      experimentally. This will be a difficult problem,
      which must be clarified before further conclusions
      on the catalytic mechanism can be drawn.

(2)   The precise nature of the involved redox states of
      the enzyme-bound selenium is still a subject of
      debate. Does the seleninic acid derivative of the
      enzyme ($R-Se\,OOH$) represent an oxidized form
      occurring during catalysis ? Can the formation of
      radical intermediates, which might be of functional
      importance, be excluded?

(3)   The involvement of other amino acid side chains
      possibly acting in concert with the selenocysteine
      residue is unclear. Trp 138 and Gln 70 may be
      possible candidates forming hydrogen bonds to the
      active group and to the substrates, but their
      precise function is not understood at present.

(4)   How does selenium enter the enzyme? A post-trans-
      lational process appears likely, but recently

evidence has been obtained that a selenocysteine specific tRNA indeed occurs in rat liver.

(5) The binding stoichiometries of the cosubstrate GSH and of several inhibitors suggest half-site-reactivity for GSH peroxidase. Does the binding process reflect this behavior by the occurrence of two different binding constants for GSH?

(6) Furthermore the mechanism by which an enzyme present in the soluble fraction of cells can react with membrane-bound peroxidized lipids is not clear. Does this mechanism exist at all?

(7) As a consequence of the GSH peroxidase reaction organic hydroperoxides are metabolized to water and the corresponding alcohol. However, it has never been shown directly that within biological systems monohydroxy polyenic fatty acids are formed from peroxidized membrane lipids by the protective function of the enzyme.

A deeper insight into these problems certainly would help to understand the catalytic function of this fascinating selenoenzyme on a molecular level. It is hoped that an approach of several disciplines acting in concert, which certainly is necessary now, will throw more light on these fundamental questions.

Acknowledgements

The author would like to thank Prof. Robert Huber for continuous support. The collaboration with Dr. Otto Epp and Prof. Albrecht Wendel during the structure analysis of GSH peroxidase is greatly appreciated. Gina Beckmann prepared the manuscript with admirable patience.

REFERENCES

(1) Babior, B., Oxygen - dependent microbial killing by Phagocytes, New England J. Med., _296_, 721 (1978)

(2) Mills, G.C., Hemoglobin catabolism. I. Glutathione
     peroxidase, an erythrocyte enzyme which protects
     hemoglobin from oxidative breakdown, J. Biol. Chem.
     229, 189 (1957).

(3) Neubert, D., Wojtczak, A. B., and Lehninger, A.L.,
     Purification and enzymatic identity of mitochondri-
     al contraction factors I and II, Proc. Natl. Acad.
     Sci., USA. 48, 1651 (1962).

(4) Little, C., and O'Brien, P. J., An intracellular
     GSH peroxidase with a lipid peroxide substrate,
     Biochem. Biophys. Res. Commun. 31, 145 (1968).

(5) Christophersen, B. O., Formation of monohydroxy-
     polyenic fatty acids from lipid peroxides by a
     glutathione peroxidase, Biochim. Biophys. Acta,
     164, 35 (1968).

(6) Rotruck, J. T., Pope, A. L., Ganther, H. E.,
     Swanson, A. B., Hafeman, D., and Hoekstra, W. G.,
     Selenium: Biochemical role as a component of
     glutathione peroxidase, Science, 179, 588 (1973).

(7) Flohe, L., Guenzler, W.A., and Schock, H. H.,
     Glutathione peroxidase: A selenoenzyme, FEBS Lett.,
     32, 132 (1973).

(8) Lawrence, R. A., and Burk, R. F., Glutathione
     peroxidase activity in selenium-deficient rat
     liver, Biochem. Biophys. Res. Commun., 71, 952
     (1976).

(9) Prohaska, J. R., Glutathione peroxidase activity of
     Glutathione-S-Transferases, Biochim. Biophys. Acta,
     611, 87 (1980).

(10) Srivastava, S.K., and Beutler, E., The transport of
     oxidized glutathione from human erythrocytes, J.
     Biol. Chem., 244, 9 (1969).

(11) Sies, H., Gerstenecker, C., Menzel, H., and Flohe, L., Oxidation of the NADPH system and release of GSSG from hemoglobin-free perfused rat liver during peroxidatic oxidation of glutathione by hydroperoxides, FEBS Lett., 27, 171 (1972).

(12) Nishiki, K., Jamieson, D., Oshino, N., and Chance, B., Oxygen toxicity in the perfused rat liver and lung under hyperbaric conditions, Biochem. J., 160, 343 (1976).

(13) Sies, H., and Summer, K. H., Hydroperoxide metabolizing systems in rat liver, Eur. J. Biochem., 57, 503 (1975).

(14) Oshino, N., and Chance, B., Properties of glutathione release observed during reduction of organic hydroperoxides, demethylation of amino-pyrine and oxidation of some substances in perfused rat liver and their implications for the physiological function of catalase, Biochem. J., 162, 509 (1977).

(15) Sies, H., Bartoli, G. M., Burk, R. F., and Waydhas, C., Glutathione efflux from perfused rat liver after phenobarbital treatment during drug oxidations and in selenium deficiency, Eur. J. Biochem., 89, 113 (1978).

(16) Hafeman, D. G., and Hoekstra, W. G., Lipid peroxidation in vivo during vitamin E and selenium deficiency in the rat as monitored by ethane evolution, J. Nutr. 107, 666 (1977).

(17) Flohe, L., Glutathione peroxidase: Fact and fiction, in: Oxygen free Radicals and Tissue Damage (Ciba Foundation Symposium 65), Excerpta Medica, Amsterdam, Oxford, New York (1979).

(18) Wendel, A., Glutathione peroxidase, in: Enzymatic Basis of Detoxication, Vol. I, Academic Press, Inc. (1980).

(19) Flohe, L., Glutathione peroxidase brought into Focus, in: Free Radicals in Biology, Vol. V., Academic Press, Inc., (1982).

(20) Flohe, L., Eisele, B., and Wendel, A., Glutathione-peroxidase. I. Reindarstellung und Molekulargewichtsbestimmungen, Hoppe Seyler's Z. Physiol. Chem. 353, 987 (1971).

(21) Nakamura, W., Hosoda, S., and Hayashi, K., Purification and properties of rat liver glutathione peroxidase, Biochim. Biophys. Acta. 358, 251 (1974).

(22) Sunde, R. A., Ganther, H. E., and Hoekstra, W. G., A comparison of ovine liver and erythrocyte glutathione peroxidase, Fed. Proc., Fed. Am. Soc. Exp. Biol., 37, 757 (1979).

(23) Awasthi, Y. C., Beutler, E., and Srivastava, S. K., Purification and properties of human erythrocyte glutathione peroxidase, J. Biol. Chem., 250, 5144 (1975).

(24) Chiu, D. T. Y., Stults, F. H., and Tappel, A. L., Purification and properties of rat lung soluble glutathione peroxidase, Biochim. Biophys. Acta, 445, 558 (1976).

(25) Awasthi, Y. C., Dao, D. D., Lal, A. K., and Srivastava, S. K., Purification and properties of glutathione peroxidase from human placenta, Biochem. J., 177, 471 (1979).

(26) Holmberg, N. J., Purification and properties of glutathione peroxidase from bovine lens, Exp. Eye Res., 7, 570 (1968).

(27) Zakowski, J. J., and Tappel, A. L., Purification and properties of rat liver mitochondrial glutathione peroxidase, Biochim. Biophys. Acta, 526, 65 (1978).

(28) Wendel, A., Glutathione peroxidase, in: Methods in Enzymology, Vol. 77, 325 (1981).

(29) Flohe L., and Brand, J., Some hints to avoid pitfalls in quantitative determination of glutathione peroxidase (E. C. 1.11.1.9.) Z. Klin. Chem. Klin. Biochem., 8, 156 (1970).

(30) Oh, S. H., Ganther, H. E., and Hoekstra, W. G., Selenium as a component of glutathione peroxidase isolated from ovine erythrocytes, Biochemistry, 13, 1825 (1974).

(31) Zakowski, J. J., and Tappel, A. L., A semi automated system for measurement of glutathione in the assay of glutathione peroxidase, Anal. Biochem., 89, 430 (1978).

(32) Guenzler, W. A., Kremers, H., and Flohe, L., An improved coupled test procedure for glutathione peroxidase (E. C. 1.11.1.9) in blood, Z. Klin. Chem. Klin. Biochem., 12, 444 (1974).

(33) Paglia, D. E., and Valentine, W. N., Studies on the quantitative and qualitative characterization of erythrocyte glutathione peroxidase, J. Lab. Clin. Med., 70, 158 (1967).

(34) Guenzler, W. A., Glutathion - peroxidase. Kristallisation, Selengehalt, Aminosaeure-zusammensetzung und Modellvorstellungen zum Reaktionsmechanismus. Ph. D. Thesis, University of Tuebingen (1974).

(35) Ladenstein, R. and Wendel, A., Crystallographic Data of the selenoenzyme glutathione peroxidase. J. Mol. Biol., 104, 877 (1976).

(36) Ladenstein, R., Epp, O., Bartels, K., Jones, A., Huber, R., and Wendel, A., Structure Analysis and Molecular Model of the Selenoenzyme Glutathione Peroxidase at 2.8 Å Resolution, J. Mol. Biol., 134, 199 (1979).

(37) Flohe, L., Guenzler, W. A., and Ladenstein, R.,
     Glutathione peroxidase, in: Glutathione (Arias, J.
     M. and Jakoby, W. B., eds.) Raven Press, New York
     (1976).

(38) Flohe, L., Guenzler, W., Jung, G., Schaich, E., and
     Schneider, F., Glutathion-Peroxidase, II. Substrat-
     spezifitaet und Hemmbarkeit durch Substratanaloge.
     Hoppe Seyler´s Z. Physiol. Chem. 352, 159 (1971).

(39) Flohe, L., Schaich, E., Voelter, W. and Wendel, A.,
     Glutathion-Peroxidase, III. Spektrale Charakterist-
     ika und Versuche zum Reaktionsmechanismus.
     Hoppe-Seyler´s Z. Physiol. Chem. 352, 170 (1971).

(40) Epp, O. Ladenstein, R. and Wendel, A., The Refined
     Structure of the Selenoenzyme Glutathione
     Peroxidase at 0.2 nm Resolution. Eur. J. Biochem.
     (1983), in the press.

(41) Kraus, R. J. and Ganther, H. E., Reaction of
     Cyanide with Glutathione Peroxidase. Biochem.
     Biophys. Res. Comm. 96, 1116 (1980).

(42) Dalziel, K., Initial steady state velocities in the
     evaluation of enzyme-coenzyme-substrate reaction
     mechanisms. Acta Chem.Scand. 11, 1706 (1957).

(43) Flohe, L., Loschen, G., Guenzler, W. A., and
     Eichele, E., Glutathione peroxidase, V. The kinetic
     mechanism. Hoppe Seyler´s Z. Physiol.Chem. 353, 987
     (1972).

(44) Guenzler, W. A., Vergin, H., Mueller, J., and
     Flohe, L., Glutathion-Peroxidase, VI. Die Reaktion
     der Glutathion-Peroxidase mit verschiedenen
     hydroperoxiden. Hoppe-Seyler´s Z. Physiol.Chem.
     353, 1001 (1972).

(45) Chiu, D., Fletcher, B., Stults, F. Zakowski, J.,
     and Tappel, A.L., Properties of selenium-gluta-
     thione peroxidase, Fed. Proc. 34, 925 (1975),
     abstr. 3996.

(46) Flohe, L. and Guenzler, W. A., Glutathione
     Peroxidase. In: Glutathione (Flohe, L.,
     Benoehr, H. Ch., Sies, H., Waller, H.D. and Wendel,
     A., eds.) Thieme, Stuttgart, 132 (1974).

(47) Wendel, A., Kerner, B. and Graupe, K., The Selenium
     Moiety of Glutathione Peroxidase. In: Functions
     of Glutathione in Liver and Kidney (Sies, H. and
     Wendel, A. eds) Springer, Berlin, 107 (1978).

(48) Wendel, A., Pilz, W., Ladenstein, R., Sawatzki, G.
     and Weser, U., Substrate-induced redox-change of
     selenium in glutathione peroxidase studied by X-ray
     photoelectron spectroscopy. Biochim. Biophys. Acta,
     377, 211 (1975).

(49) Rossmann, M. G. and Blow, D. M., The Detection of
     Sub-Units Within the Crystallographic Asymmetric
     Unit. Acta Cryst. 15, 24 (1962).

(50) Ladenstein, R., Epp, O., Roemisch, A. und Wendel,
     A., Structural Studies on the Selenoenzyme GSH
     Peroxidase. In: Trace Element Metabolism in Man
     and Animals , Proceedings of the 3rd International
     Symposium (Freising) 72 (1977).

(51) Chothia, C., and Janin, J., Principles of
     protein-protein recognition. Nature (London) 256,
     705 (1975).

(52) Crawford, J. L., Lipscomb, W. N. and Schellman, C.
     G., The Reverse Turn as a Polypeptide Conformation
     in Globular Proteins. Proc. Natl. Acad. Sci. USA,
     70, 538 (1973)

(53) Soederberg, B. O., Sjoeberg, B. M., Sonnerstam, U.,
     and Braenden, C. I., Three-dimensional structure of
     thioredoxin induced by bacteriophage T4. Proc.
     Natl. Acad. Sci. USA 75, 5827 (1978)

(54) Braenden, C. I., Relation between structure and
     function of α/β - proteins. Quarterly Rev. Biophys.
     13, 317 (1980)

(55) Hol, W. G. J., van Duijnen, P. T., and Berendsen, H. J. C., The α-helix dipole and the properties of proteins. Nature <u>273</u>, 443 (1978)

(56) Condell, R. A., and Tappel, A. L., Amino acid sequence around the active site selenocysteine of rat liver glutathione peroxidase. Biochim. Biophys. Acta, <u>709</u>, 304 (1982)

(57) Ploegmann, J. H., Drent, G., Kalk, K. H., Hol, W. G. J., Heinrikson, R. L., Keim, P., Weng, L., and Russell, J., The covalent and tertiary structure of bovine liver rhodanese. Nature, <u>273</u>, 124 (1978)

(58) Hawkes, W. C., Lyons, D. E., and Tappel, A. L., Identification of a selenocysteine-specific aminoacyl transfer RNA from rat liver. Biochim. Biophys. Acta, <u>699</u>, 183 (1982)

(59) Young, P. A., and Kaiser, I. I., Aminoacylation of Escherichia coli cysteine tRNA by selenocysteine. Arch. Biochem. Biophys. <u>171</u>, 483 (1975)

(60) Chance, B., Boveris, A., Nakase, Y., and Sies, H., hydroperoxide Metabolism: An Overview. In: Functions of Glutathione in Liver and Kidney (H. Sies and A. Wendel eds.) Springer-Verlag, Berlin, Heidelberg, New York, p. 95 (1978)

(61) Lazdunski, M., Flip-flop mechanisms and half-site enzymes. Curr. Topics Cell. Regul. <u>6</u>, 267 (1972)

THE STRUCTURE AND FUNCTION OF NEURAMINIDASE

PETER M. COLMAN
CSIRO Division of Protein Chemistry
Parkville, Victoria, Australia

ABSTRACT

Neuraminidases, originally discovered in viruses and bacteria, have captured increasing attention in recent years with their widespread discovery in animal and, in particular, mammalian tissues. Deficiencies of the enzyme in humans are now firmly associated with certain genetic disorders. The best character- ised neuraminidases are those of microbial origin. Sequences of several influenza neuraminidases are known and the three dimensional structure of one of these has recently been deter- mined. This review focusses on that structure and its relation- ship to the occurrence and biological roles of neuraminidase, its catalytic properties and its physical and chemical structure.

1. INTRODUCTION

Neuraminidase is unique among enzymes in that it was first discovered on a virus. Hirst (1), when studying the agglutination of red blood cells by influenza virus, noticed that the cells would eventually separate, whereas fresh cells could be agglutinated by the eluted virus. He correctly ascribed the transformation in the cells to an enzyme action of the virus.

215

Burnet and Stone (2) subsequently characterised an enzyme from
culture filtrates of *Vibrio cholerae* which, like the putative
influenza enzyme, destroyed receptors on red cells and rendered
them non-agglutinable by influenza virus. Hirst (1) originally
proposed that the hemagglutination of cells was a manifestation
of the formation of an enzyme-substrate complex which led to a
catalytic modification of the substrate and dissociation of the
complex. For influenza, and other orthomyxoviruses, this picture
is too simplistic. Separate proteins, a haemagglutinin and a
neuraminidase mediate the two viral functions of attachment and
receptor destruction (3,4,5). For paramyxoviruses, however the
two functions are found on one polypeptide, the HN protein (6)
although whether one or two separate active sites are responsible
is still not clear (7).

The early observations of Hirst and Burnet were followed up
by Gottschalk who observed the activity of the enzyme on muco-
proteins (8) and subsequently demonstrated that it was a gly-
cosidase (9), cleaving the glycosidic linkage between terminal
N-acetylneuraminic acid and α-linked sugars (10,11). Thus the
pursuit of an unknown enzyme activity led to the discovery of the
product, N-acetylneuraminic acid (sialic acid) a hitherto unknown
component of glycoproteins, proteoglycans and glycolipids. Sialic
acid is now recognised to be of central importance in many biological systems (12).

Here we shall continue to use the name 'neuraminidase'
however inappropriate that may be for an enzyme hydrolysing sialic
acid glycosides. Sialidase (13) is favored by some authors;
proposals to rewrite the nomenclature in a more systematic way
(14), whereby neuraminidase would be known as sialosidase, have
much to recommend them.

Earlier reviews have been written on viral and bacterial
neuraminidases (15), neuraminidase as a tool in structure
analysis (16), neuraminidase of influenza virus (17,18), altered

behaviour of neuraminidase treated mammalian cells (19) and
neuraminidases in general (20).

2. OCCURRENCE AND BIOLOGICAL ROLES

Neuraminidases are found in viruses, bacteria, protozoa and
mammals.   There are still no reports of this enzyme in the plant
kingdom.

Among viruses, only the myxoviruses, i.e. those with affinity
for certain mucins, contain neuraminidase.   In paramyxovirus
(parainfluenza, mumps, Newcastle Disease) haemagglutinin and
neuraminidase activity both reside on one protein, HN (6).
Orthomyxoviruses such as influenza express these activities on
different proteins, although one of the influenza neuraminidases
exhibits haemagglutinating activity at low (4°C) temperature (21).
The viral enzyme is always expressed as an integral membrane
protein.   It can be solubilised either by mild proteolysis (3,4)
or treatment with detergents (5).   The soluble product of mild
pronase digestion of whole influenza virus carries the full
enzymatic and antigenic capability of the membrane associated
neuraminidase (22) and in some cases is crystallisable.

Bacterial neuraminidases are commonly soluble, and are
secreted, often in large quantities, into the culture medium (2).
Exceptions include *Corynebacterium diphtheriae, Pasteurella
multocida* (23) and *Klebsiella aerogenes* (24) where the neuraminidase is membrane bound.   Bacterial neuraminidases are
inducible enzymes, and the product, N-acetylneuraminic acid, is
an inducer.   Flashner et al. (25) have recently reported that of
a number of neuraminic acid derivatives investigated, the most
effective inducer of *Arthrobacter sialophilus* neuraminidase is
the transition state analog 2,3 dehydro-N-acetylneuraminic acid.

A neuraminidase activity under developmental control has
recently been reported for the protozoan parasite *Trypanosoma
cruzi* (26).

The detection of neuraminidase in mammalian tissues has been widespread in recent years. The list includes plasma (27), mammary gland, brain, liver, kidney, small intestine, spleen, testes (28), urinary tract (29,30), heart muscle (31), eyes (32, 33), leucocytes (34,35), placenta (36) and skin fibroblasts (37). It is likely that this list will continue to expand. Although some soluble forms have been reported, notably in liver (38,39) and mammary glands (40) nearly all of the known occurrences are in association with membrane. It may be that the soluble forms are proteolytic products of the membrane forms, recalling that influenza neuraminidase can be solubilized by protease without loss of activity (22).

Finally, a neuraminidase with endo activity has recently been reported in a bacteriophage (41). The substrate for this enzyme is the host capsule polysaccharide, colominic acid, a homopolymer of $(\alpha,2{\rightarrow}8)$ linked N-acetylneuraminic acid residues.

The biological roles of different neuraminidases are many and varied. The microbial enzyme is doubtless essential for the efficacy of the virus or bacterium expressing it. A general role of conferring mobility on myxoviruses seems most likely for the viral enzyme. Here the enzyme may be important in transporting the virus through mucin to susceptible cells (16) and permitting the elution and dispersal of progeny virions from infected cells (42). A similar role for the enzyme may be found in bacterial infections where sialic-acid-like molecules act as inducers (24). The biology of the bacterial enzyme, however, is only poorly understood. In mammals where the distribution of neuraminidase through different tissues is extensive the biological functions vary widely. They include fertilisation, blood clot formation, neurotransmission, catabolism and cellular transformation (20). Certain genetic disorders, sialodoses, are typified by a deficiency of neuraminidase (43,34,35).

3. CATALYTIC PARAMETERS

3.1. Specificity

All known neuraminidases are characterised by their ability
to hydrolyse α-o-ketosidically bound neuraminic acid derivatives.
The enzyme is inactive on β-anomers. Most α-ketosidic linkages,
$(\alpha, 2 \to 3)$, $(\alpha, 2 \to 4)$, $(\alpha, 2 \to 6)$ and $(\alpha, 2 \to 8)$ are susceptible (15)
although individual enzyme preparations typically exhibit
restricted linkage specificity.

Viral neuraminidases, in particular from Newcastle disease
virus, Asian influenza and fowl plague virus, preferentially
cleave $(\alpha, 2 \to 3)$ linkages (44,45,46). Higher cleavage rates are
sometimes a result of increased V_{max} and sometimes because of
reduced K_m. In a study of viral and bacterial neuraminidases it
was found (45) that core oligosaccharide or even the protein
structure of a glycoprotein may influence the rate of hydrolysis
of sialic acid at different linkages.

Most bacterial neuraminidases show a broad range of
specificity. *Vibrio cholerae* and *Clostridium perfringens* neuraminidases (15) cleave a range of $(\alpha, 2 \to 3)$ and $(\alpha, 2 \to 6)$ linkages.
On the other hand there are examples of restricted specificities
such as that shown by neuraminidase from Type III Group B
Streptococci (47). There, not only is the enzyme specific for
the $(\alpha, 2 \to 6)$ linkage, but also for N-acetyl galactosamine adjacent
to sialic acid. Group A Streptococcal neuraminidase shows
similar specificity (48). Two neuraminidases from *Arthrobacter*
ureafaciens hydrolyse N-glycolyl - as well as N-acetyl-neuraminic
acid (49) from $(\alpha, 2 \to 3)$, $(\alpha, 2 \to 6)$ and $(\alpha, 2 \to 8)$ linkages.

A broad spectrum of linkage specificities has also been
reported for mammalian neuraminidases. Spermatozoal neuraminidase
preferentially cleaves at $(\alpha, 2 \to 6)$ linkage (50) but the fibroblast
and leucocyte enzyme prefer $(\alpha, 2 \to 3)$ linked substrates (35). Liver

(51), and heart neuraminidase (31) cleave mucus glycoprotein and gangliosides more effectively than fetuin or ($\alpha,2\rightarrow3$) linked neuraminyl lactose.

Only one endo-neuraminidase has been reported (41) and that is in association with a bacteriophage where the specificity is for the ($\alpha,2\rightarrow8$) linkage of sialic acid in colomic acid.

3.2. Inhibitors

This topic is well reviewed for influenza neuraminidases by Bucher and Palese (18). The list of inhibitors includes phosphotungstic acid, sulphydryl reagents, phenyl-gloxal derivatives and derivatives of deoxyneuraminic acid. Of the latter, 2-deoxy-2, 3-dehydro-N-trifluoroacetylneuraminic acid is the most potent with a K_i of 8×10^{-7}M (52). This compound inhibits replication of influenza in tissue culture (53). The neuraminidase activity of the HN protein of paramyxoviruses is inhibited non-competitively by high concentrations of halide ions (54).

Specific labelling of tryptophan residues in influenza neuraminidase abolishes activity either by direct interference with the catalytic site or an indirect effect on the integrity of the protein structure (55). The former interpretation is more likely in light of the three dimensional structure (see section 5.3.6). Chemical labelling of tyrosine or cysteine on the other hand, had no effect on enzyme activity (55).

EDTA inhibits some bacterial neuraminidases (15). This inhibition may usually be overcome by the addition of calcium ions. Deoxyneuraminic acid derivatives are also inhibitors of the bacterial enzyme (56).

Bacterial and viral neuraminidases show product inhibition by N-acetylneuraminic acid with K_i of the order of 1mM (15).

Little is known about inhibition of mammalian neuraminidases. One interesting observation, however, is that 20mM calcium ion inhibits spermatozoal neuraminidase (50). This is in striking

contrast to the behaviour of some bacterial neuraminidases and
indeed of liver neuraminidase which is stimulated by calcium ions.
Human placental neuraminidase is inhibited by 2-deoxy-2,3-didehydro-
N-acetylneuraminic acid with K_i of 1.6μM (36).

4. PHYSICAL CHARACTERISTICS

The neuraminidase of influenza is a mushroom-shaped pro-
jection on the surface of the virus. Electron micrographs of
detergent solubilised (57) and protease solubilised (58)
neuraminidase show a head domain of 4 subunits in a square planar
arrangement (80x80x40Å) attached to a long (∿100Å) thin stalk (59)
by means of which the head is held in association with the viral
membrane (Fig. 1). The molecular weights of the tetrameric protease released heads and the intact detergent solubilised enzyme
are 200kD and 240kD respectively (58,60). Carbohydrate is
attached to both the stalk and the head domains of the neuraminidase (61), with approximately half of the sugar content being
found in the stalk in some strains of influenza (62).

The influenza neuraminidase is oriented with its N-terminal
region in the viral membrane (60), unlike the influenza haemagglutinin and the majority of membrane bound proteins where the
C terminal domain is associated with the membrane (see e.g. 63).

Treatment of whole influenza virus with pronase liberates
soluble heads of neuraminidase as a result of cleavage at or near
residue 74 (60). These heads, which behave as tetramers of 50 kD
subunits in the ultracentrifuge (60) possess the full enzymatic
and antigenic capability of the membrane associated enzyme (22).

In contrast to the viral neuraminidases, the bacterial protein shows no evidence of subunit structure. The polypeptide
chain sizes are typically between 50kD and 100kD, although smaller
values of 39kD, for one of the two neuraminidases of *A. ureafaciens*
(49) and 10-20kD, for some estimates of *V. cholerae* neuraminidase
(64) have also been reported. Whether or not these low values

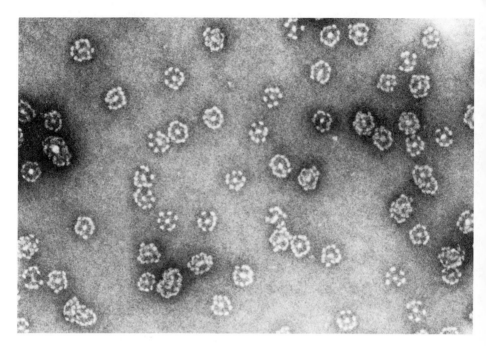

FIG. 1   Electron micrographs of
          (a) SDS solubilised influenza neuraminidase
          showing rosettes of tetrameric neuraminidase held
          together by interactions between the hydrophobic
          sequences normally associated with the viral membrane
          (b) pronase released influenza neuraminidase
          showing isolated tetrameric heads. Rectangular shapes
          are side views.
          (Micrographs by and courtesy of N. Wrigley).

represent enzymatically active proteolytic breakdown products of
a larger polypeptide remains to be seen. More common values are
90kD and 68kD for *V. cholerae* (65,15), 56kD for *Cl. perfringens*
(66), and 70kD for *Diplococcus pneumoniae* (67,68).

     The soluble forms of neuraminidase observed in bacterial
culture filtrates stand in marked contrast to the membrane bound
forms found in viruses and most mammalian cells. Solubilisation
of human liver neuraminidase by Triton X-100 (51) led to protein

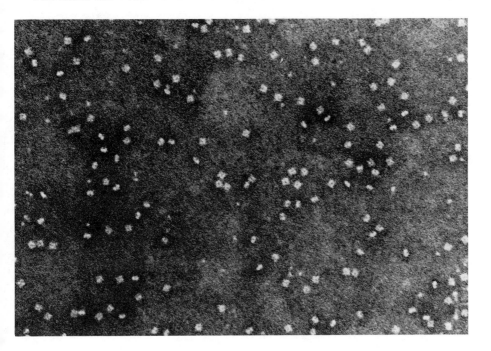

FIG. 1 cont.

molecular weights of 70kD by electrophoresis and 200kD by gel
filtration, although not all of the neuraminidase activity was
solubilised from the membrane in these experiments. Two bio-
chemically distinguishable components of human leucocyte
neuraminidase have molecular weights of 203kD and 240kD as determined by radiation inactivation, a technique that does not require
solubilisation (34). Similarly, two distinguishable components
of human placental neuraminidase were shown to have molecular
weights of 67kD and 63kD (36). This observation has led these
same authors to propose that the leucocyte enzyme is a tetramer
with subunit molecular weight similar to that observed for the
placental enzyme. Although the evidence is far from definitive,
the groupings of molecular weight estimates of mammalian

neuraminidases into the 60-70kD and 200-240kD ranges point to a
similarity with the properties of the influenza enzyme which is
difficult to overlook.

5. INFLUENZA NEURAMINIDASE

5.1. Antigenic Variation

The neuraminidase and haemagglutinin of influenza both
exhibit antigenic variation of two types (69). Minor changes
(drift) in antigenicity have now been associated with the
accumulation of point mutations in the influenza gene segments
coding for neuraminidase and haemagglutinin and usually result in
some serological cross reactivity between parent and derivative
viral strain (69,70,71). Major changes (shift) in antigenic
structure are most likely caused by the replacement of the entire
gene segment coding for either or both of the haemagglutinin and
the neuraminidase with a gene from the animal reservoir of
influenza viruses (69,70,71). These changes in subtype occur
at irregular intervals. The serological manifestation of such
a change is no cross-reactivity between antigens of different
subtypes.

Since influenza was first isolated from humans in 1933, the
following subtype changes in the haemagglutinin (H) and neuraminidase (N) have occurred:

        1933 - 1956   H1 N1
        1957 - 1967   H2 N2 (Asian flu)
        1968 - 1976   H3 N2 (Hong Kong flu)
        1977 - present   H1 N1 (Russian flu)

Three different types of influenza, A,B and C, are known.
Only the A viruses exhibit antigenic shift. Influenza C differs
from its counterparts in having only a single surface glycoprotein, with no α-neuraminidase activity (72).

5.2. Amino Acid Sequences, Disulphide Bonds and Carbohydrate Composition

The only neuraminidase sequences known at present are from influenza viruses. Eight complete sequences are available, two from N1 subtypes (63,73), five from N2 subtypes (74-78) and one from an influenza B (79). It is beyond the scope of the article to discuss antigenic variation within subtypes. This topic will be dealt with elsewhere (80). Here, we shall concentrate on three prototype sequences representative of the three subtypes sequenced so far, viz. A/PR/8/34 (N1, (63)), A/RI/5/57 (N2, (74)) and B/Lee/40 (79). An alignment of the N1, N2 and B neuraminidase sequences is given in Table 1.

Partial sequence data from the amino-terminal end of other neuraminidase subtypes (81) extend and confirm the impression from this table that the N-terminal region of sequence is unusually variable between subtypes. Among influenza A neuraminidase sequences the first six residues are invariant and believed to be located on the cytoplasmic side of the membrane. Then follows a span of 29 hydrophobic amino acids which are presumably associated with the lipid bilayer. The putative transmembrane sequence of influenza haemagglutinin is similar in length, although located in the carboxy-terminal region of the sequence (82). It is likely that the neuraminidase and haemagglutinin transmembrane segments are helical (83). The following 40 amino acids of N2 neuraminidase show no homology with other subtypes. This sequence from the membrane to the beginning of the globular neuraminidase head, which is liberated by pronase, must build the stalk of the mushroom shaped molecule visible in electron micrographs. The conformation of the glycosylated stalk peptide is the subject of some speculation. Cysteine residues provide an opportunity for disulphide bonding the four chains in pairs. If the stalk is indeed unstructured as the sequence data suggest, the carbohydrate

TABLE 1

Amino acid sequence alignment for N1 (A/PR/8/34), N2 (A/RI/5⁻/57) and B (Lee/40) strain influenza virus neuraminidases.

```
N2  Met Asn Pro Asn Gln Lys  Thr  Ile Thr Ile Gly Ser  Val Ser  Leu |15
N1  Met Asn Pro Asn Gln Lys  Ile  Ile Thr Ile Gly Ser  Ile Cys  Leu
B   Met Leu Pro Ser Thr Val Gln  -  Thr  Leu Thr Leu Leu Leu Thr
```

```
N2  Thr Ile Ala Thr Val Cys Phe Leu Met  Gln Ile Ala Ile Leu Ala |30
N1  Val Val Gly Leu Ile Ser Leu Ile Leu  Gln Ile Gly Asn Ile Ile
B   Ser Gly Gly Val Leu Leu Ser Leu Tyr Val Ser Ala Ser Leu Ser
```

```
N2  Thr Thr Val Thr Leu His Phe Lys Gln His Glu Cys Asp Ser Pro |45
N1  Ser Ile Trp Ile Ser His Ser Ile Gln Thr Gly Ser Gln Asn* His
B   Tyr Leu Leu Tyr Ser Asp Val Leu Leu Lys Phe Ser Ser Thr Lys
```

```
N2  Ala Ser Asn Gln Val Met Pro Cys Glu Pro Ile Ile Ile Glu Arg |60
N1  Thr Gly Ile Cys Asn Gln Asn Ile Ile Thr Tyr Lys Asn* Ser Thr
B   Thr Thr Ala Pro Thr Met Ser Leu Glu Cys Thr Asn Ala* Ser Asn
```

```
N2  Asn* Ile Thr Glu Ile Val Tyr Leu Asn* Asn* Thr Thr Ile Glu Lys |75
N1  Trp Val  -   -   -   -   -   -   -   -   -   -   -   -   -
B   Ala Gln Thr Val Asn* His Ser Ala Thr Lys Glu Met Thr Phe  -
```

```
N2  Glu Ile Cys Pro Glu Val Val Glu Tyr Arg Asn* Trp Ser Lys Pro |90
N1   -   -  Lys Asp Thr Thr Ser Val Ile Leu Thr Gly Asn* Ser Ser
B    -   -   -  Pro Pro Pro Glu Pro Glu Trp Thr Tyr Pro Arg Leu
```

```
N2  Gln Cys Gln Ile Thr Gly Phe Ala Pro Phe Ser Lys Asp Asn Ser |105
N1  Leu Cys Pro Ile Arg Gly Trp Ala Ile Tyr Ser Lys Asp Asn Ser
B   Ser Cys Gln Phe Gln Lys Ala Leu Leu Ile Ser Pro His Arg Phe
        <Gly Ser Thr>
```

```
N2  Ile Arg Leu Ser Ala Gly Gly Asp Ile Trp Val Thr Arg Glu Pro |120
N1  Ile Arg Ile Gly Ser Lys Gly Asp Val Phe Val Ile Arg Glu Pro
B   Gly Glu Ile Lys Gly Asn Ser Ala Pro Leu Ile Ile Arg Glu Pro
```

```
N2  Tyr Val Ser Cys Asp Pro Gly Lys Cys Tyr Gln Phe Ala Leu Gly |135
N1  Phe Ile Ser Cys Ser His Leu Glu Cys Arg Thr Phe Phe Leu Thr
B   Phe Val Ala Cys Gly Pro Lys Glu Cys Arg His Phe Ala Leu Thr
```

TABLE 1 cont.

```
N2  Gln Gly Thr Thr Leu Asp Asn Lys His Ser Asn Gly Thr Ile His 150
N1  Gln Gly Ala Leu Leu Asn Asp Arg His Ser Asn Gly Thr Val Lys
B   His Tyr Ala Ala Gln Pro Gly Gly Tyr Tyr Asn Gly Thr Arg Lys

N2  Asp Arg Ile Pro His Arg Thr Leu Leu Met Asn Glu Leu Gly Val 165
N1  Asp Arg Ser Pro Tyr Arg Ala Leu Met Ser Cys Pro Val Gly Ala
                                                        (Glu)
B   Asp Arg Asn Lys Leu Arg His Leu Val Ser Val Lys Leu Gly Ile
                                                        (Lys)

N2  Pro Phe His Leu Gly Thr Lys Glu Val Cys Val Ala Trp Ser Ser 180
N1  Pro Ser Pro Tyr Asn Ser Arg Phe Glu Ser Val Ala Trp Ser Ala
B   Pro Thr Val Glu Asn Ser Ile Phe His Met Ala Ala Trp Ser Gly

N2  Ser Ser Cys His Asp Gly Lys Ala Trp Leu His Val Cys Val Thr 195
N1  Ser Ala Cys His Asp Gly Met Gly Trp Leu Thr Ile Gly Ile Ser
B   Ser Ala Cys His Asp Gly Arg Glu Trp Thr Tyr Ile Gly Val Asp

N2  Gly Asp Asp Arg Asn Ala Thr Ala Ser Phe Ile Tyr Asp Gly Arg 210
N1  Gly Pro Asp Asn Gly Ala Val Ala Val Leu Lys Tyr Asn Gly Ile
B   Gly Pro Asp Asn Asp Ala Leu Val Lys Ile Lys Tyr Gly Glu Ala

N2  Leu Val Asp Ser Ile Gly Ser Trp Ser Gln Asn Ile Leu Arg Thr 225
N1  Ile Thr Glu Thr Ile Lys Ser Trp Arg Lys Lys Ile Leu Arg Thr
B   Tyr Thr Asp Thr Tyr His Ser Tyr Ala His Asn Ile Leu Arg Thr

N2  Gln Glu Ser Glu Cys Val Cys Ile Asn Gly Thr Cys Thr Val Val 240
N1  Gln Glu Ser Glu Cys Ala Cys Val Asn Gly Ser Cys Phe Thr Ile
B   Gln Glu Ser Ala Cys Asn Cys Ile Gly Gly Asp Cys Tyr Leu Met

N2  Met Thr Asp Gly Ser Ala Ser Gly Arg Ala Asp Thr Arg Ile Leu 255
N1  Met Thr Asp Gly Pro Ser Asp Gly Leu Ala Ser Tyr Lys Ile Phe
B   Ile Thr Asp Gly Ser Ala Ser Gly Ile Ser Lys Cys Arg Phe Leu

N2  Phe Ile Lys Glu Gly Lys Ile Val His Ile Ser Pro Leu Ser Gly 270
N1  Lys Ile Glu Lys Gly Lys Val Thr Lys Ser Ile Glu Leu Asn Ala
B   Lys Ile Arg Glu Gly Arg Ile Ile Lys Glu Ile Leu Pro Thr Gly
```

TABLE 1 cont.

N2 Ser Ala Gln His Ile Glu Glu Cys Ser Cys Tyr Pro Arg Tyr Pro 285
N1 Pro Asn Ser His Tyr Glu Glu Cys Ser Cys Tyr Pro Asp Thr Gly
B Arg Val Glu His Thr Glu Glu Cys Thr Cys Gly Phe Ala Ser Asn*

N2 Asp Val Arg Cys Ile Cys Arg Asp Asn Trp Lys Gly Ser Asn Arg 300
N1 Lys Val - Cys Val Cys Arg Asp Asn Trp His Gly Ser Asn Arg
       (Met)
B Lys Thr Glu Cys Ala Cys Arg Asp Asn Ser Tyr Thr Ala Lys Arg
       (Ile)

N2 Pro Val Ile Asp Ile Asn Met Glu Asp Tyr Ser Ile Asp Ser Ser 315
N1 Pro Trp Val Ser Phe Asp Gln Asn Asp Tyr Gln Ile Gly -   -
                               (Leu)
B Pro Phe Val Lys Leu Asn Val Glu Asp Thr Ala Glu Ile - Arg
                               (Thr)

N2 Tyr Val Cys Ser Gly Leu Val Gly Asp Thr Pro Arg Asn Asp Asp 330
N1 Tyr Ile Cys Ser Gly Val Phe Gly Asp Asn Pro Arg Pro Lys Asp
B Leu Met Cys Thr Lys Thr Tyr Leu Asp Thr Pro Arg Pro Asp Asp

N2 Ser Ser Ser Asn Ser Asn Cys Arg Asp Pro Asn Asn Glu Arg Gly 345
N1 Gly Thr -   - Gly Ser Cys Gly - Pro Val Tyr Val Asp Gly
B Gly Ser Ile Ala Gly Pro Cys Glu Ser - Asn Gly Asp Lys Trp

N2 Asn Pro Gly Val Lys Gly Trp Ala Phe Asp Asn Gly Asp Asp Val 360
N1 Ala Asn Gly Val Lys Gly Phe Ser Tyr Arg Tyr Gly Asn Gly Val
B Leu Gly Gly Ile Lys Gly Gly Phe Val His Gln Arg Met Ala Ser

N2 Trp Met Gly Arg Thr Ile Asn Lys Glu Ser Arg Ser Gly Tyr Glu 375
N1 Trp Ile Gly Arg Thr Lys Ser His Ser Ser Arg His Gly Phe Glu
B Lys Ile Gly Arg Trp Tyr Ser Arg Thr Met Ser Lys Thr Asn Arg

N2 Thr Phe Lys Val Ile Gly Gly Trp Ser Thr Pro Asn Ser Lys Ser 390
N1 Met Ile Trp Asp Pro Asn Gly Trp Thr Glu Thr Asp Ser Lys Phe
B Met Gly Met Glu Leu Tyr Val Lys Tyr Asp Gly Asp Pro Trp Thr

N2 Gln Val Asn Arg Gln Val Ile Val Asp Asn Asn Asn Trp Ser Gly 405
N1 Ser Val - Arg Gln Asp Val Val Ala Met Thr Asp Trp Ser Gly
B Asp Ser - Asp Ala Leu Thr Leu Ser Gly Val Met Val Ser Ile

TABLE 1 cont.

N2 | Tyr Ser Gly | Ile | Phe | Ser Val Glu Gly Lys^{415}
N1 | Tyr Ser Gly | Ser | Phe | Val Gln His Pro Glu (Leu Thr Gly - Leu)
B | Glu Glu Pro Gly Trp Tyr Ser Phe Gly Phe (Glu Ile Lys Asp Lys)

N2 Ser | Cys Ile | Asn Arg | Cys Phe | Tyr | Val Glu Leu Ile Arg Gly Arg^{430}
N1 Asp | Cys Ile | Arg | Pro Cys Phe | Trp | Val Glu Leu Ile Arg Gly Arg
B Lys | Cys | Asp Val | Pro Cys | Ile Gly Ile | Glu | Met Val His Asp Gly

N2 | Pro | Gln | Glu Thr | Arg Val Trp | Trp Thr Ser | Asn | Ser | Ile Val Val^{445}
N1 | Pro Lys Glu | Lys | - Thr Ile | Trp Thr Ser Ala Ser | Ser Ile Ser
B | Gly | Lys | Asp | Thr | -   -   - | Trp | His | Ser Ala | Ala Thr Ala Ile

N2 | Phe Cys Gly | Thr Ser Gly Thr Tyr Gly Thr Gly | Ser Trp Pro Asp^{460}
N1 | Phe Cys Gly | Val Asn Ser Asp Thr Val Asp | Trp Ser Trp Pro Asp
B | Tyr | Cys | Leu Met Gly Ser Gly Gln Leu Leu | Trp | Asp Thr Val Thr

N2 | Gly Ala | Asn Ile Asn | Phe | Met Pro Ile^{469}
N1 | Gly Ala | Glu Leu Pro | Phe | Thr Ile Asp Lys
B | Gly | Val Asp Met Ala Leu -   -   -

Glycosylation sequences Asn,X,Ser/Thr are marked with an asterisk.
Sequence homologies in this alignment, which is partly based on
the three dimensional structure of the N2 protein, are boxed.

may play a central role in protecting this peptide from proteolysis.

Protein sequence data on liberated neuraminidase heads
reveal two N termini corresponding to residues 74 and 77 in the
N2 sequence (75). In fact some preparations of N2 heads behave
as populations of monomers and dimers on SDS gel electrophoresis
suggesting that some cleavage with pronase may occur on the
C-terminal side of Cys 78.

Carbohydrate has been found attached to asparagine residues
86,146,200 and 234 of A/Tokyo/3/67 and N2 neuraminidase (84).

The oligosaccharide at residues 146 and 234 are of the complex type and that at 146 is further distinguished by containing N-acetylgalactosamine, an uncommon sugar residue in N-glycosidic carbohydrates. This is also the only carbohydrate attachment sequence that is conserved in N1, N2 and B neuraminidases. It has been suggested (85) that incomplete glycosylation of neuraminidase can result in the production of an inactive enzyme. Whilst this may be an indirect effect on the overall stability of the three dimensional structure of the protein, it is interesting to observe that the only conserved glycosylation site, at residue 146, is in close proximity to the enzyme active site. The triplet sequence containing carbohydrate at residue 146 is Asn-Asp-Thr in later N2 neuraminidases (75). Among the survey of Struck and Lennarz (86) all of the ten Asn-Asp-$\frac{Thr}{Ser}$ sequences were unglycosylated. Cysteine, tyrosine, glutamic acid and arginine are now the only residues that have not been observed in the second position of a glycosylated triplet (86).

One of the glycosylation triplets, that at Asn 402, has no oligosaccharide attached to it (75).

Disulphide bonds of the N2 enzyme have been determined, in part chemically and in part from the three dimensional structure (87,88). They are between residues 92-417, 124-129, 175-193, 183-230, 232-237, 278-291, 280-289, 318-317 and 421-447.

5.3.   Three Dimensional Structure of N2 Enzyme

X-ray crystallographic studies of two N2 influenza neuraminidase heads have led to a description of the three dimensional structure at 2.9Å resolution (88). The influenza neuraminidase and haemagglutinin (83) are the only membrane protein structures known at high resolution at this time. They are both subunit proteins with a single rotational symmetry axis normal to the viral membrane. Trimeric structures like the haemagglutinin are rare but tetrameric structures with four fold rotation symmetry,

like the neuraminidase (89) are unprecendented. Four fold
rotation axes have been observed in octomeric assemblies of
subunits (90,91) but never before in isolation in a tetramer.

The following discussions of the neuraminidase three
dimensional structure are based upon the interpretation of the
experimentally phased image. Not all side chain orientations are
clear at present, and discussions must be limited to those aspects
of the model which are secure.

5.3.1. Taxonomy

The backbone chain tracing of the neuraminidase has revealed
a completely new topological class of structure, a β-sheet
propeller (88). A diagram of the chain folding is shown in Fig. 2.
The polypeptide backbone passes sequentially through six four-
stranded antiparallel β sheets in anticlockwise order as the subunit is viewed into its upper face (i.e. that outermost from the
viral membrane). Each β sheet has the topology of a 'W' or
+1,+1,+1 in Richardson's (92) notation. The first strand of each
sheet to be built is that nearest the subunit centre and the last
is at the subunit surface. All sheets show the typical right-
handed twist (93) with the interior strands nearly parallel to
the molecular four fold symmetry axis and the exterior strands
inclined to be more nearly in the plane of the tetrameric assembly.
In each sheet, two of the loops connecting strands are on the upper
surface (L_{01} connecting adjacent sheets and L_{23} connecting strands
2 and 3 within sheets) and two are on the bottom surface (L_{12} and
L_{34}). Notating the jth strand of the ith sheet as $\beta_i S_j$ the folding
pattern can be described as follows:

N terminal arm on bottom of subunit

$$\beta_6 S_4$$
$$\beta_1 L_{01}, \beta_1 S_1, \beta_1 L_{12}, \beta_1 S_2, \beta_1 L_{23}, \beta_1 S_3, \beta_1 L_{34}, \beta_1 S_4$$
$$\beta_2 L_{01}, \ldots$$

$$\ldots \ldots \beta_6 S_2, \beta_6 L_{23}, \beta_6 S_3,$$
C terminal arm at subunit interface (88).

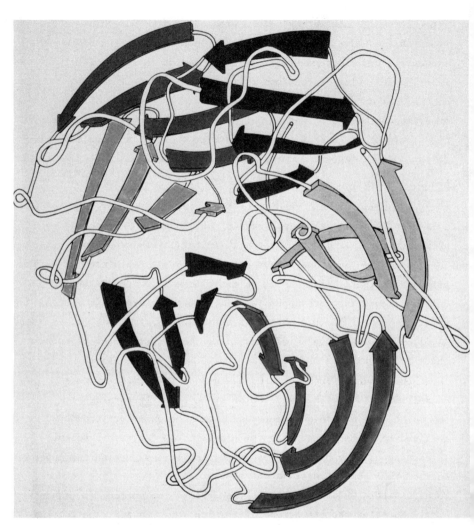

FIG. 2   Diagram of chain folding in the influenza neuraminidase
         head. The six β sheets are labelled as are the N and C
         termini. The N terminus in this diagram is residue 83,
         beyond which the interpretation of the image is
         uncertain. The four fold symmetry axis is also shown
         in the lower right hand corner.
         Reprinted with permission from Nature (Ref 88).

Antiparallel β-sheets are common structures in protein molecules. Five taxonomic subclasses have been identified to date (94)
but none of these is strictly applicable to neuraminidase. The
'Up-and-Down β barrel' observed in papain (95), has +1,+1,+1,+1,+1
topology as does part of the 'Greek key barrel' of the immuno-
globulin variable domain (96). Soybean trypsin inhibitor (97)
and catalase (98) provide other examples of this topology within
a barrel structure. The Open Face β Sandwich in thermolysin (99),
glyceraldehyde phosphate dehydrogenase (100) and the bacteriochlorophyll protein (101) all include segments of four or more
antiparallel strands connected by reverse turns. In all cases
these antiparallel structures are only part of a more extensive
secondary structure.

There are nine intrachain disulphide bonds in the neuraminidase subunit (88) and their distribution raises the question of
classifying neuraminidase as a collection of small disulphide-
rich domains (94). Figure 3 shows diagramatically the disulphide
connections between and within sheets of the N2 neuraminidase
subunit. Four of the five intra sheet disulphides link strands
one and two (in sheets 1,3 and 4). None of the other covalent
links show topological similarities to each other. Although
most of the proteins in the 'Small Disulphide-rich' class are
single domain structures (94), the wheat germ agglutinin displays
4 topologically identical domains of about 40 amino acids each
with 4 disulphide bonds (102). However the β sheet topology is
quite different to the 'W' found in neuraminidase.

Only among fibrous proteins has there been a report of a
four stranded antiparallel β sheet (103) with +1,+1,+1 topology.
The evidence for the topology, however is tenuous being based on
schemes of predicting conformation from sequence (104).

Most significantly, the neuraminidase fold does not belong
to the 'swiss-roll' β structure class categorising a number of

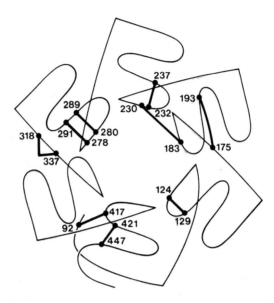

FIG. 3   Diagram of the disulphide bonding in influenza
          neuraminidase N2 heads.

lectins, (94) including the influenza haemagglutinin (83) which
is a sialic acid specific lectin. Apparently the two distinct
functions involving terminal sialic acid residues, namely their
recognition and their enzymatic cleavage, have evolved
independently. This may not always be the case, as evidenced by
the haemagglutinating activities of some neuraminidases (6,21).

      Evidently the family of 'Antiparallel β domains' (94)
needs to be expanded to include the β sheet propeller. The extent
of the interactions between the six sheets in the neuraminidase
structure (see below) requires that it be considered as a single
domain structure of approximately 390 amino acids. Other large
single domain structures are triosephosphate isomerase (247
residues (105)), carboxypeptidase (307 residues (106)) and the
bacteriochlorophyll protein (358 residues (101)).

5.3.2. β-Sheet Folding

Each of the six sheets contains three reverse turns, L_{12}, L_{23} and L_{34} and one turn connecting it to the preceding sheet, L_{01} (88). These loop structures are variable in length and they carry amino acids implicated in enzyme activity and antigenic variation (107). The conservation of small amino acids 164 (gly, $\beta_1 L_{34}$) 177 (ala or gly, $\beta_2 L_{01}$), 186 (gly, $\beta_2 L_{12}$), 196 (gly, $\beta_2 L_{23}$), 235 (gly, $\beta_3 L_{12}$) 244 (gly, $\beta_3 L_{23}$), 261 (gly, $\beta_3 L_{34}$), 270 (gly or ala, $\beta_4 L_{01}$) and 348 (gly, $\beta_5 L_{01}$) in some loops points to their importance in mediating reversal of chain direction. Other loops are bounded by disulphide bonds; in particular $\beta_1 L_{12}$, $\beta_3 L_{12}$, $\beta_4 L_{12}$ and $\beta_6 L_{23}$.

A preliminary table of secondary structure assignments has been published (88) and is the basis for figure 2. Refinement of the structure is necessary before inter-strand hydrogen bonding assignments can be made. The structure in $\beta_4 S_1$ and $\beta_4 S_2$ is especially interesting because it includes the disulphide linkages 278-291 and 280-289. Apparently disulphides may not join neighbouring strands in a β sheet (94). The backbone conformation in this region of neuraminidase is presumably relaxed from regular β strand to allow covalent linkage in this case.

5.3.3. Propeller Stabilisation

Between all pairs of sheets are found cores of hydrophobic amino acids, listed in Table 2. Many of these residues are either conserved across influenza subtypes or conservatively exchanged. Some of these cores appear to be more extensive than others. At the centre of the subunit, S_1 on all sheets run parallel to each other. Viewed from above the subunit the six strands project onto the corners of a distorted hexagon such that $\beta_2 S_1$ and $\beta_5 S_1$ are farther apart than any other pairs. The possibility of some parallel sheet structure linking β_1 to β_3 and

TABLE 2

Amino acids in the hydrophobic cores between adjacent sheets in the N2 neuraminidase propeller.

Core	Contributing Amino Acids
$\beta_1 - \beta_2$	Phe 132, Leu 134, Leu 158 Trp 189
$\beta_2 - \beta_3$	Val 192, Phe 205, Tyr 207 Met 241, Leu 255, Ile 257, Ile 262
$\beta_3 - \beta_4$	Val 240, Ile 254 Cys 278, Cys 280, Cys 289, Cys 291
$\beta_4 - \beta_5$	Ile 290 Trp 383
$\beta_5 - \beta_6$	Trp 352, Phe 354, Trp 361, Tyr 374 Phe 422, Val 424, Phe 446, Phe 97
$\beta_6 - \beta_1$	Phe 410, Tyr 423, Phe 100

β_4 to β_6 cannot yet be excluded but the chains appear too far apart for this type of interaction.

Two classes of sheet-sheet packing in globular proteins have been identified, aligned and orthogonal (108). The neuraminidase is clearly in the former category. Furthermore the geometrical arrangement of the sheets in the propeller requires that the intersheet angle be negative (109,110).

5.3.4. Subunit Interactions

The arrangement of subunits in the tetramer is shown diagrammatically in Figure 4. The single subunit interface consists primarily of residues from the outside strands (§4) of sheets β_6, β_1 and β_2, the C terminal strand and the loops L_{01}, L_{23} and L_{34} on β_1. Hydrophobic and hydrophilic interactions are found across the interface. Two possibilities are envisaged at present for continuing the sheet secondary structure across the interface. Both include $\beta_1 S_4$, in one case in conjunction with

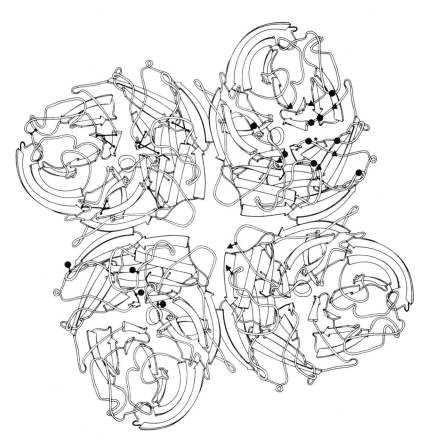

FIG. 4   Diagram of the arrangement of subunits in the tetrameric
           influenza neuraminidase head. Each subunit highlights
           different features of the structure, viz. (reading anti-
           clockwise from top left) disulphide bonds, carbohydrate
           attachment, putative metal ligands (Asp 113, Asp 141)
           and conserved acidic (circles) and basic (triangles)
           residues in the sialic acid binding site (asterisk).
           Reprinted with permission from Nature (Ref 88).

$\beta_6 S_4$ and in the other with $\beta_1 L_{34}$, a part of which may be in antiparallel β arrangement with $\beta_1 S_4$ on a neighbouring subunit. Which,
if either, of these prevails can not yet be unequivocally stated.
Around the fourfold symmetry axis, His 168 makes contacts with its
neighbours in a way which is unique to N2 structures. Nearer the
upper surface of the head, eight acidic groups (Asp 113, Asp 141
on all four subunits) cluster around a site which can be labelled
by Sm^{+++}. Calcium ions are presumed to be bound here in the native
enzyme, by anology with other systems (99) and in keeping with
observations of calcium requirements for neuraminidase stability
(111,112). One of the carbohydrate moieties, at Asn 200, appears
partly to cover the surface of a neighbouring subunit and may
thereby contribute to the stabilisation of the tetrameric head.
This carbohydrate attachment sequence, however, is not conserved
in N1 and B strains (Table 1).

5.3.5. Carbohydrate Attachment Sites

It is not yet possible to describe in detail the polypeptide
backbone conformation around the glycosylation sites of the
neuraminidase. Asparagine residues 86, 146 and 200 are in chain
segments with extended configurations whilst residue 234 is in a
bend (Figure 4). Two of the oligosaccharides (146,200) are on
the upper surface of the head and two (86 and 234) on the lower.
Sugars are attached to Asn residues in turns and extended chains
on influenza haemagglutinin (113) and in a bend on Fc fragment of
IgG1 (114).

The glycosylation triplet from Asn 402 contains no oligosaccharide (75). The chain configuration here is a turn such that
the middle residue of the triplet (Trp 403) is exposed in a way
that would hinder simultaneous recognition of Asn 402 and Ser 404
by transferases. Apparently this structure is preformed before
glycosylation can occur. The transfer of sugar to protein is
cotranslational (115,116) although in some cases transfer may
occur to the native folded polypeptide chain (116). In vitro

glycosylation of nonglycosylated sequences is possible after
denaturation (86). Clearly the three dimensional protein
structure may influence the accessibility of a glycosylation
triplet even if the nascent polypeptide chain is the recipient of
the oligosaccharide. Only 49 out of 159 tripeptide glycosylation
sequences analysed in eukaryotes were actually glycosylated (86).
In this respect, glycosylation sequences near the protein
C-terminus may be more at risk than early sequences. The distance
lag between synthesis and sugar transfer is believed to be 45-80
residues (115,117).

In the present neuraminidase image (88) only the oligosaccharide at Asn 200 is visible. The structure of the sugar at
Asn 146 is especially interesting because this is the only
conserved glycosylation site in all known neuraminidase amino acid
sequences and because of the implication, directly or otherwise
of the role of carbohydrate in the enzyme activity (85).

Influenza B strain neuraminidase has no glycosylation
sequences on the N-terminal arm on the bottom surface of the
subunit. However, a compensating sequence is found nearby on the
bottom surface at residue 285 (N2 numbering). Thus N1, N2 and B
neuraminidase structures all have two potential sites for carbohydrate attachment on the bottom face of the subunit.

5.3.6. The Enzyme Active Centre

The active site of the neuraminidase has been located by
soaking sialic acid, a known inhibitor (see above), into the
crystals and using difference Fourier methods to determine the
binding site (107). As seen looking into the top face of the
subunit, that site is almost directly above the first strands of
sheets 3 and 4 ($\beta_3 S_1$, $\beta_4 S_1$) and is located in a large pocket or
depression in the surface of the molecule (Fig. 5). Like other
enzymes catalysing the removal of end groups, rather than internal
chain cleavage, the morphology of the site is more like a pocket
than a cleft. The walls of the pocket are lined by residues from

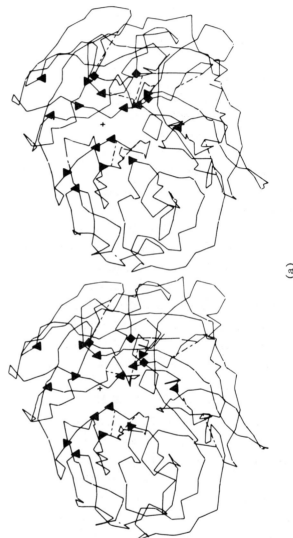

(a)

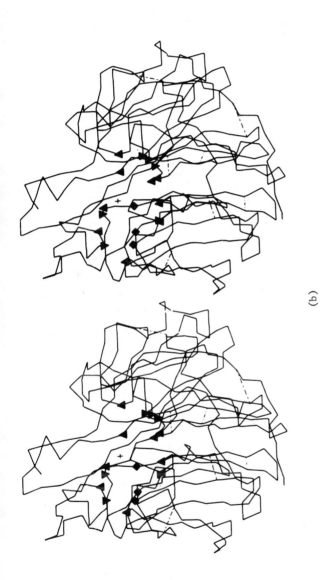

(b)

FIG. 5   Stereo diagram of the sialic acid binding site of influenza neuraminidase
          showing
          ◄ Glu 119, Asp 151, Asp 198, Glu 227, Asp 243, Glu 276, Glu 277,
            Asp 330, Glu 425
          ▼ Arg 118, Arg 152, Arg 224, His 274, Arg 292, Lys 350
          ◆ Tyr 121, Leu 134, Trp 178
          + Sialic acid binding site
          Disulphide bonds are dashed
          (a) view parallel to the molecular four fold symmetry axis
          (b) view normal to the molecular symmetry axis.

TABLE 3

Conserved amino acids in influenza neuraminidase sequences
implicated in enzyme activity by their proximity to the sialic
acid binding site.

Neutral	Basic	Acidic	Hydrophobic
Ser 179	Arg 118	Glu 119	Leu 134
	Arg 152	Asp 151	Trp 178
	Arg 224	Asp 198	(Tyr 121)
	His 274	Glu 227	
	Arg 292	Asp 243	
	Lys 350	Glu 276	
	(His 150)	Glu 277	
		Asp 330	
		Glu 425	

His 150 is Lys in N1 and B strains.
Tyr 121 is Phe in N1 and B strains.

$\beta_1 L_{01}$, $\beta_1 L_{23}$, $\beta_2 L_{23}$, $\beta_3 L_{01}$, $\beta_3 L_{23}$, $\beta_4 L_{01}$, $\beta_4 L_{23}$ and $\beta_5 L_{01}$. On
these loops are found an extraordinarily large number of conserved,
charged amino acids whose orientations are seen to be towards the
sialic acid binding site. These are shown in Table 3.

The binding of heavy metals to some of these residues during
the multiple isomorphous replacement phasing procedure gives some
indication of their reactivity. Diaminodinitroplatinum binds to
His 150, and mercury phenyl glyoxal binds to Arg 224, supporting
the observed inhibition of neuraminidase by phenylglyoxal
derivatives (see section 3.2.). Chemical
modification of tryptophan also abolishes activity (55). The role
of some sulphydryl reagents as inhibitors is presumably an indirect
one, and may result from opening up the disulphide linkages in the
stepladder between $\beta_4 S_1$ and $\beta_4 S_2$. There is so far no evidence for
a calcium ion in the active site although its presence cannot yet
be excluded. If indeed there is none, then the inhibitory role
of EDTA on enzyme activity must also be indirect, although the
putative calcium site on the fourfold symmetry axis is
approximately 30Å distant from the sialic acid binding site.

TABLE 4

Predicted and observed β structure in N2 neuraminidase.

Observed β structure in N2 neuraminidase heads		Predicted	
		Ref 121	Ref 123
$\beta_1 S_1$	120–123	114–118	114–118
$\beta_1 S_2$	130–134	129–134	129–134
$\beta_1 S_3$	157–162	155–161	157–160
$\beta_1 S_4$	170–174	173–179	173–177
$\beta_2 S_1$	178–185		
$\beta_2 S_2$	188–194	188–195	190–194
$\beta_2 S_3$	200–207	202–207	203–207
$\beta_2 S_4$	210–217		210–214
$\beta_3 S_1$	228–233	218–226,230–234	
$\beta_3 S_2$	237–242	236–242	237–242
$\beta_3 S_3$	251–259	252–257	
$\beta_3 S_4$	262–268		262–266
$\beta_4 S_1$	278–283		
$\beta_4 S_2$	286–291		287–291
$\beta_4 S_3$	296–304	302–307	302–308
$\beta_4 S_4$	307–314		316–323
$\beta_5 S_1$	351–356		350–354
$\beta_5 S_2$	359–366	360–366	
$\beta_5 S_3$	372–382	376–380	376–379
$\beta_5 S_4$	387–398	391–398	392–399
$\beta_6 S_1$	404–414		409–412
$\beta_6 S_2$	417–428	417–424	421–426
$\beta_6 S_3$	438–447	436–440,443–447	444–447
$\beta_6 S_4$	95–106	91–98	92–97

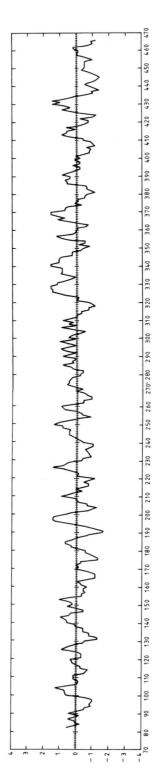

FIG. 6   Hydrophilicity profile (ordinate) of the N2 influenza neuraminidase head sequence (abscissa). Peaks around residues 153, 200, 330, 340, 370 and 430 correspond to antigenic determinants (Ref 107).

Sigmoidal kinetics have been reported for influenza (118) and human brain (119) neuraminidase. The active centres of the influenza enzyme are ~40Å apart and thus like other proteins showing interdependent ligand binding to identical subunits the effect would have to be mediated in some way through the subunit interface. No gross reorganisation around this interface occurs (120) when sialic acid binds, although the binding affinity in that case is weak (~1mM), and possibly insufficient to trigger a quaternary structure change. Further experiments are needed to clarify the results of the kinetic data.

5.3.7. Sequence and Structure

Two schemes for predicting secondary structure from sequence data have been applied to the N2 neuraminidase sequence. The Chou-Fasman (121) prediction has been given elsewhere (122) but is included here in Table 4 - for completeness together with the Robson (123) prediction. The definition of β structure in the observation list is topological rather than geometrical. For example, residues 114-118, predicted by both schemes to be in β structure, is indeed in an extended conformation which may be antiparallel β with residues 135-139. A model refinement will allow more precise definitions of the β sheet segments in the structure.

In Figure 6 is shown the hydrophilicity index which it is claimed (124) predicts potential antigenic sites on the structure. Insofar as chain segments 153, 197-199, 328-336, 339-347, 367-370, 400-403 and 431-434 are implicated in antibody binding sites on the neuraminidase (107) this scheme has some merit.

6. COMPARISON OF INFLUENZA N2, N1 and B NEURAMINIDASES

The amino acid sequence alignment in Table 1 is based, in part, on the three dimensional structure of the N2 protein. We shall not discuss the stalk and membrane peptides here (residues

1-73) except to observe that the strictly invariant N-terminal
hexapeptide in all A type neuraminidases, N1-N8 (81), is altered
in influenza B. What, if any, role this sequence may play in
determining viable reassortments of the influenza gene segments
is unclear, although it is likely that this hexapeptide is on the
cytoplasmic side of the viral membrane.

Within the globular head sequences, residues 74-469, there
are many features pointing to similarity of structure for the three
neuraminidase subtypes. Firstly, the overall level of homology,
as shown in Table 5 is sufficient to direct folding into similar
topologies. The lower level of homology between the A and B
strain sequences is largely due to differences over the 100
C terminal residues where N2 and B have only 8 amino acids in
common, and N1 and B only 13. All of the disulphide bands of the
head region are in alignment. N2 sequences have one additional
pair of Cysteine residues bridging residues 175 and 193. An
additional Cysteine in the B sequence at residue 251 (N2 numbering) is remote from the four fold axis and cannot be involved in
disulphide formation. Clusters of amino acids conserved in all
three sequences are found within the segments 146-148, 177-186,
222-237, 242-248, 276-280 and 289-294 (107). The last four
segments are presumably critical features of β sheets 3 and 4.
The upper surface loops of these sheets are rich in residues
implicated in catalytic activity (see above).

TABLE 5

Number of identical residues in pairwise comparison of N2, N1 and
B neuraminidase head sequences.

	N2	N1	B
N2	–	168 (.43)	101 (.26)
N1	–	–	116 (.29)
B	–	–	–

Fractional homology in 395 amino acids is shown in parenthesis.

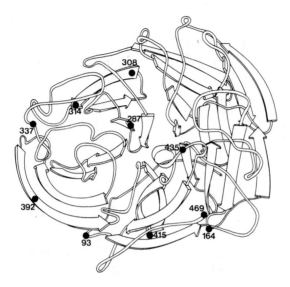

FIG. 7   Diagram of N2 influenza neuraminidase showing sites of
          insertion or deletion in N1 or B neuraminidase sequences.

The insertions and deletions found in N1 and B structures
relative to N2 are most commonly located in surface loops and are
shown in Figure 7.

(i)   Three residues are inserted into the B sequence at
position 93.   The homology in the following segment with N1 and
N2 is low but insertion here rather than elsewhere in the next
heptapeptide allows the additional structure to extend the
N-terminal arm prior to its entry into $\beta_6 S_4$.

(ii) N1 and B both have an additional charged residue in
$\beta_1 L_{34}$.   Its role is unclear.

(iii) $\beta_4 S_2$ (287) has an insertion and a deletion in N1 and
an insertion in B.   In N2, Arg 288 may be in a salt link with
Asp 304.   In the B structure the same residues may be linked,
Glu 288-Lys 304.   The salt bridge is not present in N1, where 288
is deleted and residue 304 is Serine.   On the opposite side of β_4,
N1 and B build a more extensive interface with β_5 than does N2.

The inserted residue 287a in these structures might contact Phe 238 or Tyr 238 on β_3, at which point N2 has threonine.

(iv)   N1 and B structures insert one residue at position 308 at the start of $\beta_4 S_4$.

(v)   At the end of $\beta_4 S_4$, 314 is deleted in B and 314 and 315 are deleted in N1.

(vi)   The disulphide bridge between Cys 318 and Cys 337 in $\beta_5 L_{01}$ may assist in the stabilisation of a structure whose sequence varies considerably within subtypes. Deletions in N1 and B are found either side of Cys 337.

(vii) $\beta_5 S_4$ demonstrates a β bulge (125) at residue Asn 392. N1 and B have a deletion at this point.

(viii)   N1 and B insert four and five residues respectively in to $\beta_6 L_{12}$ after residue 415.

(ix)   One and three deletions respectively are found in N1 and B at residue 435. This loop, $\beta_6 L_{23}$, represents the most distal extent of the structure from the viral membrane.

(x)   Slight variations in length are found at the C-terminus.

In summary, the structures of the N1 and B neuraminidases will almost certainly belong to the family of β-sheet propellers. Furthermore they will also display the more subtle features of this structure, e.g. disulphide bonding, sheet-sheet stacking, lengths of loops and residue orientation within the catalytic centre.

REFERENCES

1.   Hirst, G.K., J. Exp. Med., 76, 195 (1942).

2.   Burnet, F.M. and Stone, J.D., Aust. J. Exp. Biol. Med. Sci., 25, 227 (1947).

3.   Mayron, L.W., Robert, B. and Winzler, R.J., Arch. Biochem. Biophys., 92, 475 (1961).

4.   Noll, H., Aoyagi, T. and Orlando, J., Virology, 18, 154 (1962).

5. Laver, W.G., Virology, 20, 251 (1963).

6. Scheid, A., Caliguiri, L.A., Compans, R.W. and Choppin, P.W.,
    Virology, 50, 640 (1972).

7. Portner, A., Virology, 115, 375 (1981).

8. Gottschalk, A. and Lind, P.E., Nature, 164, 232 (1949).

9. Gottschalk, A., Biochim. Biophys. Acta, 20, 560 (1956).

10. Gottschalk, A., Biochim. Biophys. Acta, 23, 645 (1957).

11. Blix, G., Gottschalk, A. and Klenk, E., Nature, 179, 1088
    (1957).

12. Rosenberg, A. and Schengrund, C.-L., The Biological Roles
    of Sialic Acid, Plenum, New York, 1976.

13. Heimer, R. and Meyer, K., Proc. Nat. Acad. Sci. USA., 42,
    728 (1956).

14. Scott, J.E., Yamashina, I. and Jeanloz, R.W., Biochem. J.,
    207, 367 (1982).

15. Drzeniek, R., Curr. Top. Microbiol. Immun., 59, 35 (1972).

16. Gottschalk, A. and Drzeniek, R., The Glycoproteins,
    A. Gottschalk, Elsevier, Amsterdam, 1972, 381-402.

17. Laver, W.G., Adv. Virus Res., 18, 57 (1973).

18. Bucher, D. and Palese, P., The Influenza Viruses and
    Influenza, E.D. Kilbourne, Academic, New York, 1975, 83-123.

19. Ray, P.K., Appl. Microbiol., 21, 227 (1977).

20. Rosenberg, A. and Schengrund, C.-L., The Biological Roles
    of Sialic Acid, Plenum, New York, 1976, 295-357.

21. Laver, W.G. et al. In preparation, 1983.

22. Laver, W.G., Virology, 86, 78 (1978).

23. Drzeniek, R., Scharman, W. and Balke, E., J. Gen. Microbiol.,
    72, 357 (1972).

24. Pardoe, G.I., Path. Microbiol., 35, 361 (1970).

25. Flashner, M., Kassler, J. and Tannenbaum, S.W.,
    J. Bacteriology, 151, 1630 (1982).

26. Pereira, M.E., Science, 219, 1444 (1983).

27. Warren, L. and Spearing, C.W., Biochem. Biophys. Res. Comm.,
    3, 489 (1960).

28. Carubelli, R., Trucco, R.E. and Caputto, R., Biochim.
    Biophys. Acta, 60, 196 (1960).

29. Candiotti, A., Ibanez, N. and Monis, B., Experienta, 28, 541,
    (1972).

30. Omichi, K. and Ikemata, T., J. Chromatog., 230, 415 (1982).

31. Tallman, J.F. and Brady, R.O., Biochim. Biophys. Acta, 293,
    434 (1973).

32. Tulsiani, D.R.P., Nordquist, R.E. and Carubelli, R., Exp.
    Eye Res., 15, 93 (1973).

33. Dreyfus, H., Preti, A., Harth, S., Pellicone, C. and
    Virmaux, N., J. Neurochemistry, 40, 184 (1983).

34. Nguyen Hong, V., Beauregard, G., Potier, M., Bélisle, M.,
    Mameli, L., Gatti, R., and Durand, P., Biochim. Biophys.
    Acta, 616, 259 (1980).

35. Kuriyama, M., Someya, F., Miyatake, T. and Koseki, M.,
    Biochim. Biophys. Acta, 662, 220-225 (1981).

36. McNamara, D., Beauregard, G., Nguyen Hong, V., Yan, D.L.S.,
    Bélisle, M. and Potier, M., Biochem. J., 205, 345 (1982).

37. Beauregard, G., Melancon, S.B., Dallaire, L. and Potier, M.,
    Biochim. Biophys. Acta, 706, 212 (1980).

38. Tulsiani, D.R.P. and Carubelli, R., J. Biol. Chem., 245,
    1821 (1970).

39. Tulsiani, D.R.P. and Carubelli, R., Biochim. Biophys. Acta,
    284, 257 (1972).

40. Tulsiani, D.R.P. and Carubelli, R., Biochim. Biophys. Acta,
    227, 139 (1971).

41. Kwiatkowski, B., Boschek, B., Thiele, H. and Stirm, S.,
    J. Virology, 43, 697 (1982).

42. Palese, P., Tobita, K., Ueda, M. and Compans, R.W., Virology, 61, 397 (1974).

43. Lowden, J.A. and O'Brien, J.S., Am. J. Hum. Genet., 31, 1 (1979).

44. Corfield, A.P., Wember, M., Schauer, R. and Rott, R., Eur. J. Biochem., 124, 521 (1982).

45. Corfield, A.P., Higa, H., Paulson, J.C. and Schauer, R., Biochim. Biophys. Acta, 744, 121 (1983).

46. Paulson, J.C., Weistein, J., Dorland, L., van Halbeck, H. and Vliegenthart, J.F.G., J. Biol. Chem., 257, 12734 (1982).

47. Milligan, T.W., Mattingly, S.J. and Straus, D.C., J. Bacteriol., 144, 164 (1980).

48. Davis, L., Baig, M.M. and Ayoub, E.M., Infect. Immun., 24, 780 (1979).

49. Uchida, Y., Tsukada, Y. and Sugimori, T., J. Biochem., 86, 1573 (1979).

50. Srivistava, P.N. and Abou-Issa, H., Biochem. J., 161, 193 (1977).

51. Michalski, J.-C., Corfield, A.P. and Schauer, R., Hoppe-Seyler's Z. Physiol. Chem., 363, 1097 (1982).

52. Meindl, P., Bodo, G., Palese, P., Schulman, J. and Tuppy, H., Virology, 58, 457 (1974).

53. Palese, P., Schulman, J.L., Bodo, G. and Meindl, P., Virology, 59, 490 (1974).

54. Merz, D.C., Prehm, P., Scheid, A. and Choppin, P.W., Virology, 112, 296 (1981).

55. Bachmeyer, H., Febs Lett., 23, 217 (1972).

56. Meindl, P. and Tuppy, H., Hoppe-Seyler's Z. Physiol. Chem., 350, 1088 (1969).

57. Laver, W.G. and Valentine, R.C., Virology, 38, 105 (1969).

58. Wrigley, N.G., Skehel, J.J., Charlwood, P.A. and Brand, C.M., Virology, 51, 525 (1973).

59. Wrigley, N.G., Brit. Med. Bull., 35, 35 (1979).

60. Blok, J., Air, G.M., Laver, W.G., Ward, C.W., Lilley, G.G., Woods, E.F., Roxburgh, C.M. and Inglis, A.S., Virology, 119, 109 (1982).

61. Klenk, H.-D. and Rott, R., Curr. Top. Microbiol. Immun., 90, 19 (1980).

62. Allen, A.K., Skehel, J.J. and Yuferov, V., J. Gen. Virol., 37, 625 (1977).

63. Fields, S., Winter, G. and Brownlee, G.G., Nature, 290, 213 (1981).

64. Rosenberg, A., Binnie, B. and Chargraff, E., J. Am. Chem. Soc., 82, 4113 (1960).

65. Laver, W.G., Pye, J. and Ada, G.L., Biochim. Biophys. Acta, 81, 177 (1964).

66. Balke, E. and Drzeniek, R., Z. Naturforsch., 24, 599 (1969).

67. Stahl, W.L. and O'Toole, R.D., Biochim. Biophys. Acta, 268, 480 (1972).

68. Tannenbaum, S.W. and Sun, S.-C., Biochim. Biophys. Acta, 229, 824 (1971).

69. Webster, R.G. and Laver, W.G., The Influenza Viruses and Influenza, E.D. Kilbourne, Academic, New York, 1975, 269-314.

70. Laver, W.G. and Air, G.M., Structure and Variation in Influenza Viruses, Elsevier, New York, 1980.

71. Laver, W.G., The Origin of Pandemic Influenza Viruses, Elsevier, New York, 1983.

72. Herrler, G., Nagele, A., Meier-Ewert, H., Bhown, A.S. and Compans, R.W., Virology, 113, 439 (1981).

73. Hiti, A.L. and Nayak, D.P., J. Virol., 41, 730 (1982).

74. Elleman, T.C., Azad, A.A. and Ward, C.W., Nucleic Acids Res., 10, 7005 (1982).

75. Ward, C.W., Elleman, T.C. and Azad, A.A., Biochem. J., 207, 91 (1982).

76. Markoff, L. and Lai, C.-J., Virology, 119, 288 (1982).

77. Bentley, D.R. and Brownlee, G.G., Nucleic Acids Res., 10, 5033 (1982).

78. Van Rompuy, Min-Jou, W., Huylebroeck, D. and Fiers, W., J. Mol. Biol., 161, 1 (1982).

79. Shaw, M.W., Lamb, R.A., Erikson, B.W., Breidis, D.J. and Choppin, P.W., Proc. Nat. Acad. Sci. USA., 79, 6817 (1982).

80. Colman, P.M. and Ward, C.W., Curr. Top. Microbiol. Immun., In preparation.

81. Blok, J. and Air, G.M., Biochemistry, 21, 4001 (1982).

82. Ward, C.W., Curr. Top. Microbiol. Immun., 59, 35 (1981).

83. Wilson, I.A., Skehel, J.J. and Wiley, D.C., Nature, 289, 366 (1981).

84. Ward, C.W., Murray, J.M., Roxburgh, C.M. and Jackson, D.C., Virology, 126, 370 (1983).

85. Griffin, J.A., Basak, S. and Compans, R.W., Virology, 125, 324 (1983).

86. Struck, D.K. and Lennarz, W.J., The Biochemistry of Glyco-proteins and Proteoglycans, W. Lennarz, Plenum, New York, 1980, 35-83.

87. Ward, C.W., Colman, P.M. and Laver, W.G., Febs Lett., 153, 29 (1983).

88. Varghese, J.N., Laver, W.G. and Colman, P.M., Nature, 303, 35 (1983).

89. Colman, P.M. and Laver, W.G., Structural Aspects of Recognition and Assembly in Biological Macromolecules, M. Balaban, ISS, Rehovot, 1981, 869-872.

90. Ward, K.B., Hendrickson, W.A. and Klippenstein, G.L., Nature, 257, 818 (1975).

91. Stenkamp, R.E., Sieker, L.C., Jensen, L.H. and McQueen, J.E. Jnr., Biochemistry, 17, 2499 (1978).

92. Richardson, J.S., Nature, 268, 495 (1977).

93. Chothia, C., J. Mol. Biol., 75, 295 (1973).

94. Richardson, J.S., Adv. Prot. Chem., 34, 167 (1981).

95. Drenth, J., Jansonius, J.N., Koekoek, R. and Wolthers, B.G., Adv. Prot. Chem., 25, 79 (1971).

96. Schiffer, M., Girling, R.L., Ely, K.R. and Edmundson, A.B., Biochemistry, 12, 4620 (1973).

97. Sweet, R.M., Wright, H.T., Janin, J., Chothia, C. and Blow, D.M., Biochemistry, 13, 4212 (1974).

98. Vainshtein, B.K., Melik-Adamyan, V.R., Barynin, V.V. and Vagin, A.A., Dokl. Akad. Nank. S.S.S.R., 250, 242 (1980).

99. Colman, P.M., Jansonius, J.N. and Matthews, B.W., J. Mol. Biol., 70, 701 (1972).

100. Buehner, M., Ford, G.C., Moras, D., Olsen, K.W. and Rossman, M.G., J. Mol. Biol., 90, 25 (1974).

101. Matthews, B.W., Fenna, R.E., Bolognesi, M.C., Schmid, M.F. and Olson, J.M., J. Mol. Biol., 131, 259 (1979).

102. Wright, C.S., J. Mol. Biol., 111, 439 (1977).

103. Fraser, R.D.B., MacRae, T.P., Parry, D.A.D. and Suzuki, E., Polymer, 12, 35 (1971).

104. Fraser, R.D.B. and MacRae, T.P., Proc. 16th Int. Ornithological Congr., H.J. Frith and J.S. Calaby, Griffin, Adelaide, 1976, 443-451.

105. Banner, D.W., Bloomer, A.C., Petsko, G.A., Phillips, D.C., Pogson, C.I. and Wilson, I.A., Nature, 255, 609 (1975).

106. Quiocho, F.A. and Lipscomb, W.N., Adv. Prot. Chem., 25, 1 (1971).

107. Colman, P.M., Varghese, J.N. and Laver, W.G., Nature, 303, 41 (1983).

108. Chothia, C., Levitt, M. and Richardson, D.C., Proc. Nat. Acad. Sci. USA., 74, 4130 (1977).

109. Chothia, C. and Janin, J., Proc. Nat. Acad. Sci. USA., 78, 4146 (1981).

110. Cohen, F.E., Sternberg, M.J.E. and Taylor, W.R., J. Mol. Biol., 148, 253 (1981).

111. Baker, N.J. and Gandhi, S.S., Archs. Virol., 52, 7 (1976).

112. Carroll, S.M. and Paulson, J.C., Archs. Virol., 71, 273 (1982).

113. Wilson, I.A., Ladner, R.C., Skehel, J.J. and Wiley, D.C., Biochem. Soc. Trans., 11, 145 (1982).

114. Deisenhofer, J., Biochemistry, 20, 2361 (1981).

115. Rothman, J.E. and Lodish, H.F., Nature, 269, 775 (1977).

116. Bergman, L.W. and Kuehl, W.M., Biochemistry, 17, 5174 (1978).

117. Hubbard, S.C. and Ivatt, R.J., Ann. Rev. Biochem., 50, 555 (1981).

118. Mountford, C.E., Grossman, G., Holmes, K.T., O'Sullivan, W.J., Hampson, A.W., Raison, R.L. and Webster, R.G., Molec. Immunol., 19, 811 (1982).

119. Tettamanti, G., Cestaro, B., Venerendo, B. and Preti, A., Enzymes of Lipid Metabolism, S. Gatt, L. Freysz and P. Mandel, Plenum, New York, 1978, 417-437.

120. Matthews, B.W. and Bernhard, S.A., Ann. Rev. Biophys. Bioeng., 2, 257 (1973).

121. Chou, P.Y. and Fasman, G.D., Ann. Rev. Biochem., 42, 251 (1978).

122. Azad, A.A., Elleman, T.C., Laver, W.G. and Ward, C.W., Origin of Pandemic Influenza Viruses, W.G. Laver, Elsevier, New York, 1983, 59-76.

123. Garnier, J., Osguthorpe, D.J. and Robson, B., J. Mol. Biol., 120, 97 (1978).

124. Hopp, T.P. and Woods, K.R., Proc. Nat. Acad. Sci. USA., 78, 3824 (1981).

125. Richardson, J.S., Getzoff, E.D. and Richardson, D.C., Proc. Nat. Acad. Sci. USA., 75, 2574 (1978).

INSTRUCTIONS FOR THE PREPARATION OF MANUSCRIPTS
FOR DIRECT REPRODUCTION

Peptide and Protein Reviews is a book in the English language devoted to the publication of definitive review articles related to significant emerging areas of peptide and protein research. Review articles on all branches of research with amino acids, peptides and proteins, including works on their isolation, analysis, structure, biosynthesis, immunology, metabolism and other biological properties will be considered for publication. The scope of this book will be broad, extending from such basic experimental areas as peptide synthesis through to studies on the analysis of molecular function. In addition, hypothesis germane to these fields will be considered for publication. A major aim of the journal is to encourage wider awareness of important developments in methodology, and the further advancement in our understanding of the role of peptides and proteins in the Life Sciences. Most reviews are written at the invitation of a member of the International Editorial Advisory Board. Unsolicited reviews, commentaries, and hypotheses are, however, welcomed. In this case, a brief outline accompanied by key references should be sent to a member of the Board.

Directions for Submission

One typewritten manuscript suitable for direct reproduction, carefully inserted in a folder, and two (2) copies of the manuscript must be submitted. Since all contributions are reproduced by direct photography of the manuscripts, the typing and format instructions must be strictly adhered to. Noncompliance will result in return of the manuscript to the authors and delay its publication. To avoid creasing, manuscripts should be placed between heavy cardboards and securely bound before mailing. Authors are responsible for obtaining written permission from the publisher for figures, tables, or any material from previously published sources. Permission documents must be submitted to the Editor before the manuscript can be published.

Manuscripts should be mailed to:

M. T. W. Hearn
St. Vincent's School of Medical Research
Victoria Parade
Melbourne, Victoria, 3065
AUSTRALIA

Typing Instruction and Format of Manuscript

Compositor: Please try to use an IBM Selectric typewriter.

Trim Size: The average book measures 6 x 9 inches (15.25 x 22.75 cm). The image area in which you should type measures 5¾ x 8¾ inches (14.5 x 22 cm). This copy will be photographically reduced after typing.

Typefaces*: We have chosen Prestige Elite-12 for text; Dual Gothic-12 for title, author(s) name, affiliations, and headings. *If not available, please use comparable alternatives.

Paragraph Indent: Set a tab after five spaces and use throughout. However, no paragraph indent after any section headings.

Spacing: Set typewriter to 1½ line vertical spacing throughout text and single (1-line) space for your title, author and affiliation, lists, tables, and figure legends, and references.

Chapter Opening Page (first page of chapter):

Title: Dual Gothic-12: All capital letters. Flush left, do not indent. Type title 1¾ inches from the top of the image area. Do not break words; do not exceed a width of 4 inches (10.25 cm) from left margin. If title goes to a second or third line, align it with the line above (flush left, do not indent).

Author's Name(s) and Affiliation(s): Dual Gothic-12: Use capital and lower case letters (e.g. Percival N. Blakeney) flush left. At single spacing return carriage twice. After name, insert slash then insert affiliation; use capital and lower case letters. You may type to the right margin. Single space and flush left any affiliations continuing more than one line.

Note: Two or more authors having the same affiliation should be set up in the same fashion i.e., grouped with an "and" between the names, followed by a slash and affiliation written once (e.g. Percival N. Blakeney and Eleanor Guienne/). If there are two authors with two different affiliations, leave 1½ lines of space between them.

Text: Prestige Elite: Text on chapter opening page begins 4½ inches from the top of the image area; do not indent first paragraph. On all other pages, text begins on first line of image area. Text ends at 8¾ inches from the top of the image area.

Headings:

Primary (Major) Headings — Dual Gothic, all capital letters, flush left. Double space above and below headings to and from surrounding text.

Secondary Headings — Dual Gothic, capitalize the first letter of every main word; flush left. Underline entire heading. Double space above and below heading to and from surrounding text.

Tertiary Headings — Dual Gothic, capitalize the first letter of every main word flush left. Double space above and 1½ space below heading to and from surrounding text.

Numbered Lists: Prestige Elite: Double space above and below lists. Paragraph indent; type Arabic number; period; leave two spaces to text. Capitalize first letter of the first word of each entry. Text may continue to right margin. When entries exceed margin, align the next line under the first word after number. Single space within and between entires.

For Sublists: Indent lower case letters under first word after Arabic number; insert period. Leave 2 spaces to text. Text may continue to right margin. When entries exceed margin, align the next line under first word after the lower case letter. Single space within and between entries.

Unnumbered Lists: Prestige Elite; Double space above and below list. Paragraph indent; Capitalize first letter of the first word of each entry. Text may continue to right margin. When entries exceed margin indent next line two spaces from line above it. Next entry should align with first word of new entry above. Single space within and between entries.

Tables: Prestige Elite. Each table should be typed on a separate sheet of paper not as part of the text. No table should exceed one page.

Table Title: The word TABLE should be capitalized and positioned flush left above the table. Leave one space, insert Arabic numeral, period, leave two spaces to table title. The table title should have the first letter of all main words in capitals. Titles should be typed single space. If the table title exceeds the width of the table the second line should align with "T" from "TABLE" above. Leave 1½ line spacing to table body.

Table Body: Insert a rule along the width of the table. Leave one line of space to table columns. Columns Heads: Initial caps; centered over columns. Leave one line of space; insert rule along the width of the table. Leave 1½ lines of space to column entries. Single space within entries. If entry exceeds column width indent two spaces from entry above. Leave 1½ line space. Insert end rule along the width of the table.

Table Footnote: Table footnotes should be indicated by lower case superscript letters in the body of the table. Leave one line of space from table end rule to footnote. The footnote should be positioned flush left on the width of the table. The footnote should begin with the corresponding lower case superscript letter. Capitalize first word of footnote, type on full width of table, end footnote with a period.

Figures: Formulas, graphs, and other numbered figures should be professionally drawn in black India ink (do not use blue ink) preferably with Leroy lettering on separate sheets of white paper placed at the end of text. Figure should not be placed within the body of the text since the printer will reduce and position figures after receipt of manuscript. They should be sized to fit within the width and/or height of the type page, including any legend, label, or number associated with them. Photographs should be used for all labels on the figures or photographs; they may not be hand drawn. All figure legends should be typed on a separate sheet of paper. PLEASE PUT AUTHOR(S) NAME(S), ARTICLE TITLE, AND FIGURE NUMBERS ON THE BACK OF EACH FIGURE.

Figure Legend: Use Dual Gothic, the word FIGURE should be capitalized and positioned flush left with blue line. Leave on space, insert Arabic numeral, period, leave two spaces to legend. The first letter of the legend should be capitalized, followed by all lower case letters. Legend ends with a period. Legends should be typed single spaced. Type legend to right margin. If legend should exceed right margin, flush left the next line.

References: Type the word "REFERENCES" as a Primary heading. References are to be single spaced. Reference number 1-9 leave one space. Type Arabic number, period. Reference number 10 and on, flush left, period. Leave two spaces to wording align the next line under first letter of first word after number use 1½ lines of space between entries. The reference list, consecutively numbered in order of citation in the text, should be typed single spaced, although separated from one another by an extra line of space. Use Chemical Abstracts abbreviations for journal titles. References to journal articles should include (1) surname, first and middle initials, (2) journal, (3) volume number, (4) first page, and (5) year, in that order. References to books should include (1) author; surname, first and middle initials, (2) title of book, (3) editor of book (if applicable), (4) edition of book (if any), (5) publisher, (6) city of publication, (7) year of publication, and (8) page reference (if applicable). E.g., Journals: Evin, A. B. and Seyferth, D., "The Processing of Coal," J. Amer. Chem. Soc., 89, 952 (1967). Books: Pressman, D. and Grossberg, A. L., The Structural Basis of Antibody Specificity, W. A. Benjamin, Inc., New York, 1968.

Note: No footnotes should be shown at the bottom of the pages. Footnotes are to be collected and listed in the section labeled FOOTNOTES at the rear of the manuscript. They are to be indicated in the text as superscripted Arabic numerals.

Each page of manuscript should be numbered with a light blue pencil. Do not number pages with figure legends. Make sure to number the pages outside of the margin area.

It is essential to use black typewriter ribbon (carbon film ribbon is preferred) in good condition so that a clean, clear impression of the letters is obtained. Erasure marks, smudges, creases, etc., may result in return of the manuscript to the authors for retyping.